AF564870

# Artificial Seeds Technology

# About the Author

**Prof. Bidhan Roy** is teaching in the Department of Seed Science and Technology, Uttar Banga Krishi Viswavidyalaya, Pundibari, Cooch Behar, West Bengal. He actively involves himself in research, teaching of UG & PG students. Prof Roy has about 18 years of research experience in Plant Breeding, Biotechnology and Seed Production of field crops, rice in particular. He is an excellent Rice Breeder of this University and developed a number of advanced lines; few of them are in AICRIP trials, and two varieties have been identified for release. He has also an adequate amount of experienced in educating farming community through extension works. He has guided many clusters of farmers for production of quality seed of field crops and the performance of the cluster in respect of seed production is excellent. He successfully handled four research projects and some other projects are running successfully.

He is life member of a number of national and international scientific societies and Founder Secretary of Cooch Behar Association for Cultivation of Agricultural Sciences, UBKV, Pundibari, W.B. He is also serving as Technical Advisor of three Farmers' Groups who involve in seed production, agricultural technology dissemination, agricultural inputs distribution etc. He has organized many training programmes for farmers, acted as resource person in training organized by different institutes, acted as external examiners of different Agricultural Universities, reviewed research papers for different reputed journals.

He has published more than 60 original research papers of national and international reputes, nineteen book chapters in edited books on Plant Breeding and Biotechnology and he has authored three books and edited one book. To recognize his distinction he has been awarded Young Scientist Award by Crop and Weed Science Society, Bidhan Chandra Krishi Viswavidyalaya, W.B. and Fellow of Hind Agri-Horticultural Society, Muzaffarnagar 251 001, Uttar Pradesh; Senior Scientist Award by National Environmentalist Association, Ranchi University, Ranchi, Jharkhand; Scientist of the Year-2014 by Scientific and Environmental Research Institute, 42-Station Road, Rahara, Kolkata; Science Excellence Award, Foundation for Science and Environment, Kolkata; Bharat Gaurav Award, by Indian International Friendship Society, Ranjit Studio, 7-Tansen Marg; Bharat Vikas Award 2016, by Institute of Self Reliance, Plot No: 103/B, HIG, BDA Duplex, Baramunda, Bhubaneswar etc.

# Artificial Seeds Technology
## An Emerging Avenue of Seed Science and Applied Biotechnology

**Bidhan Roy**
Professor
Department of Seed Science and Technology
Uttar Banga Krishi Vishwavidayalaya
Pundibari, Cooch Behar 736 165
West Bengal, India

**NEW INDIA PUBLISHING AGENCY**
**New Delhi-110 034**

**NEW INDIA PUBLISHING AGENCY**
101, Vikas Surya Plaza, CU Block, LSC Market
Pitam Pura, New Delhi 110 034, India
Phone: + 91 (11)27 34 17 17 Fax: + 91(11) 27 34 16 16
Email: info@nipabooks.com
Web: www.nipabooks.com

Feedback at feedbacks@nipabooks.com



ISBN 978-93-87973-59-6



Composed and Designed by NIPA

*Dedicated to my*
*Patron*
**Shri Samir Acharya**

# Preface

Nowadays, artificial seed technology is one of the most important tools to breeders and scientists of plant tissue culture. It has offered powerful advantages for large scale mass propagation of elite plant species and conservation of plants. The artificial seed technology is anrousing and fast growing area of research in plant tissue culture. This technology is currently considered as an efficient alternative technique for plant propagation of commercially important plant materials. This technology also facilitates the way of handling cells and tissues, protecting them from external gradients, short-term and long-term storage under low temperature and ultra-low temperature, respectively and as an efficient system of delivery. The synthetic seed technology provides many other advantages which have been discussed in detail in different chapters in this endeavour.

The information in the areas of synthetic seed preparation technology, its implications, achievements and limitations are lying unorganized in different articles of journals and edited books. Those information was presented in this monograph in organized way with up-to-date citations, which will provide comprehensive literatures of recent advances.To be an expert in artificial seeds, one should have enough practical hand in plant tissue culture. So, in Chapter 2, Fundamentals of Plant Tissue Culture has been elaborated in simple approach. The base materials for production of synthetic seeds are somatic embryos or tissue culture derived materials, viz. shoot tips, axillary buds, meristem etc. To produce large number of synthetic seeds, a continuous supply of source material is very essential. The different sources of encapsulating plant materials have been detailed in this book as a separate chapter. A separate chapter also has been included on hardening of artificial seed derived plantlets or plant tissue cultured derived plantlets to make the *in vitro* developed plantlets suitable for planting in field condition. The main aim of artificial seeds is to obtain true-to-type plants; consequently thedifferent methods of testing clonal fidelity have been described in Chapter 9.

It is my pleasure to acknowledge the help of Ms. Priyanka Sharma and Ms. Pallabi Saha for their precarious review to improve the manuscript. I also like to acknowledge the service received from my students Mr. Gadge Susant Sundarrao and Mr. Monish Roy during preparation of the manuscript.

I am greatly obliged to Prof. Asit Kumar Basu (BCKV, Mohanpur, Nadia, West Bengal, India) and Dr. Asit B. Mandal under whose guidance I learnt the production technology of synthetic seed. I also acknowledge the encouragement and assistance received from the staff of the Department of Seed Science and Technology (UBKV, Pundibari, Cooch Behar, West Bengal) to write this book. I am indebted to the Vice-Chancellor of Uttar Banga Krishi Viswavidyalaya for allowing me in preparation of this book. I am especially indebted to my beloved son KINJAL for his many small pieces of assistance during the preparation of the manuscripts.

Any suggestion from the readers related to different aspects relevant to this book for further improvement will be highly cherished.

**Bidhan Roy**

Dated: 22nd July, 2019
Cooch Behar, India

# Contents

# Glossary

**Abiotic stresses:** Abiotic stresses such as drought (water deficit), excessive watering (water-logging/flooding), extreme temperatures (cold, frost and heat), salinity (sodicity) and mineral (metal and metalloid) toxicity negatively impact growth, development, yield and seed quality of crop and other plants.

**Acclimatization:** The process that leads to the adaptation of an organism or idiotype to a new environment.

**Additive:** Something that is added to synthetic medium/ gelling agents to improve the chemical of physical properties of medium/ gelling agent.

**Adsorption:** Capability of a solid substance (adsorbent) to attract to its surface molecules of a gas or solution (adsorbate) with which it is in contact.

**Adventitious shoot:** Shoot arising from other than the usual place.

**Aforestation:** Planting of forestry plants on non-forest area.

**Agar:** A polysaccharide bearing gel forming ability obtained from certain species of red algae.

**Albino plant:** Plant without any pigment. Many regenerated plantlets are albino, particularly most of the anther/pollen grain derived plantlets.

**Androgensis:** Development of haploid embryo from male gametophytes.

**Angiosperm:** A group of plants whose ovules (seeds) are borne within a mature ovary (fruit).

**Anther:** The pollen bearing portion of the stamen of flower.

**Antibiotic:** A drug used to treat infections caused by bacteria and other microorganisms. Originally, an antibiotic was a substance produced by one microorganism that selectively inhibits the growth of another. Synthetic antibiotics, usually chemically related to natural antibiotics, have since been produced that accomplish comparable tasks.

**Antitranspirants:** Antitranspirants are compounds applied to the leaves of plants to reduce transpiration. They are used on cut flowers, on newly transplanted shrubs, and in other applications to preserve and protect plants from drying out too quickly.

**Apical meristem:** Meristematic cells at the apex of the shoot or root.

**Atmosphere:** The mixture of gasses that surrounds the earth. *OR* The air in or around a place.

**Atomic weight:** Atomic weight of an element is the relative weight of the atom. The IUPAC definition (IUPAC, 1980) of atomic weight is 'An atomic weight (relative atomic mass) of an element from a specified source is the ratio of the average mass per atom of the element to 1/12 of the mass of an atom of $^{12}C$'.

**Autoendroreduplication:** Spontaneous chromosome doubling through anther culture of some crops (rice) is termed as autoendroreduplication.

**Autotrophic:** The characteristics of plants that they are capable of manufacturing their own food.

**Auxotroph:** A mutant organism that will not grow on minimal medium but requires the addition of some growth factor(s).

**Axillary bud:** Shoot bud formed in the axil of the leaf and the stem.

**Biotic stress:** It is the **stress** that occurs as a result of damage done to an organism by other living organisms, such as bacteria, viruses, fungi, parasites, beneficial and harmful insects, weeds, and cultivated or native plants.

**Callus induction medium:** Synthetic plant tissue culture medium used for callus induction from cultured explant is termed as callus induction medium. Commonly, it is fortified with high concentration of auxins, namely 2,4-D, 2,4,5-T, picloram, dicamba etc.

**Callus maintenance medium:** The callus maintenance medium is used for callus proliferation and it contains ½ strength of the callus inducing hormone(s).

**Callus:** A mass of actively dividing non-organized parenchyma cells derived out of tissue culture in the presence of plant hormone(s). Callus is either homogenous parenchymatus mass or treachery elements or sieve elements or submerized cells or secretory cells or the trichomes. These undifferentiated cells have the ability to dedifferentiate into organ or into a complete plant under appropriate medium supplemented with plant growth regulator(s).

**Carbohydrates**: Carbohydrates include sugars found naturally in foods such as fruits, vegetables, milk, and milk products. It is also known as saccharides or carbs, are sugars or starches. They are a major food source and a key form of energy for most organisms. They consist of carbon, hydrogen, and oxygen atoms.

**Caulogenesis:** Initiation of only adventitious shoot bud from callus is known as caulogenesis.

**Cell line:** A cell progeny/population arising from a single cell of primary suspension culture. All the cells in this population are with same morphological and physiological characters.

**Chlorophyll**: It is any member of the most important class of pigments involved in photosynthesis, the process by which light energy is converted to chemical energy through the synthesis of organic compounds. Chlorophyll is found in virtually all photosynthetic organisms, including green plants, prokaryotic blue-green algae (cyanobacteria), and eukaryotic algae. It absorbs energy from light; this energy is then used to convert carbon dioxide to carbohydrates.

**Clone:** A group of identical organism, cells, or descended from a single ancestral organism, organ, tissue or cell.

**Clonal fidelity:** True-to-type tissue culture derived plantlets or clonal propagated planting materials are termed as clonal fidelity. It is one of the most important pre-requisites in micro-propagation of crop species.

**Concentration:** Concentration of a solution is the relative proportion of the solute in relation to solvent.

**Cotyledon:** Seed-leaf storing the food in dicot plants.

**Cross-pollination:** Pollen grains from flowers of one plants pollinate the flower of other plant.

**Cryopreservation:** Cryopreservation refers to the stepwise viable freezing of biological materials (seed, planting materials, plant callus, somatic embryos, synthetic seeds etc.) followed by storage at ultra-low temperature, preferably at that of liquid nitrogen.

**Cryoprotectant:** An agent able to prevent freezing and thawing damage to cells as they are frozen or defrosted. These substances have high water solubility and low toxicity.

**Dedifferentiation:** Reversion of differentiated cells to non-differentiated cells. That is, the phenomenon of a mature reverting to a meristematic state and forming undifferentiated callus is termed as dedifferentiation.

**Differentiation:** Developmental change of a cells or tissues which leads to its performing specialized function and/or regeneration of organ (roots or shoots) or organ-like structure (pro-embryos).

**Dihaploid:** A haploid which arising from a tetraploid ($2n = 4x$), or haploid derived from tetraploid. The chromosome number in the dihaploid is $2x$.

**Disinfection:** Killing or removing of microorganisms from explant.

**Distilled water:** Water produced by distillation of containing no organic compounds.

**Dormancy:** Dormancy in a general sense is the inactive period of seed embryo which has attained maturity up to when it begins to germinate. Seed dormancy in specific sense is the condition in a viable seed by which it prevents germination. It is a special mechanism of repression of all regeneration activities for germination of seeds.

**Double haploid:** Homozygous diploid obtained through chromosome (self-chromosome doubling or colchicine induced doubling) doubling of haploid.

**EDTA (ethylenediamine tetracetic acid):** A chelating agent, used to check iron from precipitating in solution.

**Electrophoresis:** It is a molecular technique of separation of charged molecules (e.g. DNA, RNA, proteins etc.) on the basis of relative migration in an appropriate matrix (e.g. agarose, polyacylamide etc.) subjected to an electric field.

**Embriods:** Embryo like structures derived as consequence of different process such as embryogenesis or androgenesis.

**Embryo:** A young sporophytic plant, as yet retained in the gametophyte in the seed.

**Embryogenesis:** The process by which an embryo develops from a fertilized egg cell or asexually from a group of cell(s).

**Endosperm:** The triploid nutritive tissue produced within the embryo-sac of plant seeds, particularly all monocots, with some exception in dicot seeds.

**Enzyme:** A protein that acts as catalyst to speed up chemical reaction.

**Equivalent weight:** The equivalent weight is the weight of a substance that will combine with or replace one mole of hydrogen or one-half mole of oxygen. The equivalent weight is equal to the atomic weight divided by the valence.

**Essential elements:** The nutrient elements which are indispensable or necessary for the normal growth and development of plants and they are known as essential elements.

***Ex situ* conservation:** Ex situ conservation literally means, 'off-site conservation' It is the technique of conservation of all levels of biological diversity outside their natural habitats through different techniques like zoo, captive breeding, aquarium, botanical garden, and gene bank.

**Explant:** Any part or tissue of plant used to initiate *in vitro* tissue culture. It may be a piece of leaf, meristem tip, shoot tip, axillary bud, node, inter node, immature inflorescence, anther, ovule, pollen grains etc.

**Fertilization:** Union of male and female gametes in sexual reproduction.

**Gel electrophoresis:** It is a method for separation and analysis of macromolecules (DNA, RNA and proteins) and their fragments, based on their size and charge.

**Genetic stability:** a measure of the resistance to change, with time, of the sequence of genes within a DNA molecule or of the nucleotide sequence within a gene.

**Genetic transformation:** In molecular biology, **transformation** is the **genetic** alteration of a cell resulting from the direct uptake and incorporation of exogenous **genetic** material from its surroundings through the cell membrane(s).

**Germplasm:** Germplasm are living genetic resources such as seeds or tissues that are maintained for the purpose of plant breeding, preservation, and other research uses. These resources may take the form of seed collections stored in seed banks, trees growing in nurseries or gene banks, etc. Germplasm collections can range from collections of wild species to elite, domesticated breeding lines that have undergone extensive human selection. Germplasm collection is important for the maintenance of biological diversity and food security.

**Greenhouse:** A building of glass or plastic that is used to grow and protect tender plants.

**Growth retardants:** The term growth retarding or growth retardant is that the chemical slows cell division and cell elongation of shoot tissue and regulate plant height physiologically without formative effects.

**Gymnosperm:** An important division of the plant kingdom, being woody plants with alteration of generations, having the gametophyte retained on the sporophyte and seed produced on the surface of the sporophylls and not enclosed in an ovary.

**Haploid:** Having a single complete set of chromosomes per cell, usually a gametophyte.

**Hardening:** It is the process of exposing transplants (seedlings) gradually to outdoor conditions. It enables your transplants to withstand the changes in environmental conditions they will face when planted outside in the garden. It encourages a change from soft, succulent growth to a firmer, harder growth.

**Hardening-off:** Adapting plants to outdoor conditions by gradually withholding water, lowering the temperature, increasing light intensity, or reducing the nutrient supply. The hardening-off process conditions plants for survival when transplanted outdoors. The term is also used for gradual acclimatization to in vivo conditions of plants grown in vitro, e.g., gradual decrease in humidity. See also: acclimatization; freeliving conditions (see also weaning; acclimatization).

**Heterotrophic:** The plants which do not possess photosynthetic pigments and which are not capable of manufacturing their own organic food have to depend for their food requirements on outside and are called heterotrophic plants.

**Hill reaction:** Hill and coworkers showed that water is the source of oxygen evolved during photosynthesis. They proved this performing the following reaction:

$$2H_2O + 4[Fe(CN)_6]^{3-} \rightarrow 4[Fe(CN)6]^{4-} + 4H^+ + O_2$$

**Hormone:** A natural chemical that exerts strong controlling effects on growth, development and metabolism at very low concentrations and usually at sites other than that of its synthesis.

**Humidity:** Humidity is a term used to describe the amount of water vapor present in air. Water vapor, the gaseous state of water, is generally invisible to the human eye. Humidity indicates the likelihood for precipitation, dew, or fog to be present.

**Hybrid:** First generation (F1) from cross between two purelines, inbred, open-pollinated varieties, clones or other population that are genetically dissimilar.

***In vitro*:** Literally means 'in glass', i.e. growing of living things in artificial environment, mainly in vessels under control environment.

***In vivo*:** Growing of living things in natural condition.

**Inoculation:** Culture aseptic placement of plant explant on nutrient medium.

**Isozymes:** Isoenzymes (also called isozymes) are alternative forms of the same enzyme activity that exist in different proportions in different tissues. Isoenzymes differ in amino acid composition and sequence and multimeric quaternary structure; mostly, but not always, they have similar (conserved) structures. Their expression in a given tissue is a function of the regulation of the gene for the respective subunits. Each isoenzyme form will have different kinetic and/or regulatory properties that reflect its role in that tissue. Isoenzymes are generally identified in the clinical laboratory by electrophoresis.

**Leaf Cuticle:** A plant cuticle is a protecting film covering the epidermis of leaves, young shoots and other aerial plant organs without periderm. It consists of lipid and hydrocarbon polymers impregnated with wax, and is synthesized exclusively by the epidermal cells.

**Macronutrients:** The elements required comparatively large in quantity, that is, in concentrations more than 0.5 mmol $L^{-1}$ are known as macronutrients

**Material safety data sheet (MSDS):** MSDS is a form containing data regarding the properties of a particular substance. An important component of product stewardship and workplace safety, it is intended to provide workers and emergency personnel with procedures for handling or working with that substance in a safe manner, and includes information such as physical data (melting point, boiling point, flash point, etc.), toxicity, health effects, first aid, reactivity, storage, disposal, protective equipment, and spill handling procedures. The exact format of an MSDS can vary from source to source within a country depending on how specific is the national requirement.

**Meristem:** Undifferentiated but mitotically active cells of plant. The portion of the shoot lying distal to the youngest leaf primordium and measuring up to about 100 μm in diameter and 250 μm length is called the apical meristem (see Fig. 5.8).

**Meristemoids:** The localized meristematic cells on a callus which give rise to shoots and/ or roots are termed as meristemoids.

**Micronutrient:** Elements required in the concentrations less than 0.5 mmol $L^{-1}$ is know as micronutrients.

**Micropropagation:** Micropropagation is the practice of rapidly multiplying stock plant material to produce a large number of true-to-type progeny plants, using modern plant tissue culture methods.

**Microsatellite:** A microsatellite is a tract of repetitive DNA in which certain DNA motifs (ranging in length from 1-6 or more base pairs) are repeated, typically 5-50 times. Microsatellites occur at thousands of locations within an organism›s genome. They have a higher mutation rate than other areas of DNA leading to high genetic diversity. Microsatellites are often referred to as **short tandem repeats** (**STRs**) by forensic geneticists and in genetic genealogy, or as **simple sequence repeats** (**SSRs**) by plant geneticists.

**Microspore:** A spore that, in vascular plant, gives rise to male gametophyte.

**Molar (M) concentration:** A mole is the quantity of a substance whose weight in grams is equal to the molecular weight of the substance.

**Molecular weight:** The sum of the atomic weight of all the atoms in a molecule is termed as molecular weight.

**Morphogenesis:** Developmental pathways in differentiation which result in the formation of recognizable tissues.

**Morphology:** Morphology is a branch of biology dealing with the study of the form and structure of organisms and their specific structural features. This includes aspects of the outward appearance (shape, structure, colour, pattern, size), i.e. external morphology (or eidonomy), as well as the form and structure of the internal parts like bones and organs, i.e. internal morphology (or anatomy).

**Normal (N) solution:** One normal solution of a substance contains one equivalent or one gram equivalent weight of the substance in one liter of solution.

**Organogenesis:** It is type of morphogenesis which results in formation of organ (either shoot or root) from cell or tissue.

**Organoids:** Anomalous structures when develop during organogenesis is called organoids.

**Orthodox seeds:** Seeds of this type can be dried to moisture content of 5% or lower without lowering their viability. Most crop seeds belong to this category. Such seeds can be easily stored for their long periods, their longevity increases in response to lower humidity and storage temperature.

**Osmoticum:** Reagents that increase osmotic pressure of the medium.

**Ovary:** Enlarged basal portion of the pistil containing ovules.

**Ovule:** A rudimentary seed, containing before fertilization the female gametophyte with egg cell, all being surrounded by the nucellus and one or two integuments.

**PAGE (Polyacrylamide gel electrophoresis):** A methode of separating nuclic acid or protein molecules according to their size. The molecules migrate through the inert gel matrix under the influence of an electric field.

**Parasite:** An living organism that lives on or in another and gets it food from the host organism and gives nothing in return.

**Parts per million:** Number of share of million parts of a solution or mixture is termed as parts per million (ppm)

**PCR (Polymer Chain Reaction):** An *in vitro* procedure that amplifies enzymatically a particular DNA sequence which is flanked by two oligonucleotide (deoxy) primers that share identity to the opposite DNA strands. It involves multiple cycles of denaturation, annealing of oligo primers and synthesis of DNA using a thermostable DNA polymerase.

**Peat:** It is a heterogeneous mixture of more or less decomposed plant (humus) material that has accumulated in a water-saturated environment and in the absence of oxygen. Its structure ranges from more or less decomposed plant remains to a fine amorphic, colloidal mass.

**Percentage:** The dictionary meaning of percentage (%) is a rate or number or amount in each hundred. It may be expressed in w/w or w/v or v/v.

**Perlite:** It is an amorphous volcanic glass that has relatively high water content, typically formed by the hydration of obsidian. It occurs naturally and has the unusual property of greatly expanding when heated sufficiently.

**pH:** The negative logarithm of the hydrogen ion concentration of any solution for measuring acidity or alkalinity is known as pH.

**Phenolic compound**: An aromatic organic compound where one or more hydroxy groups are bonded directly to benzene. They are technically alcohols, but they have quite different chemical properties. Phenolics can also determine plant properties such as flavour and palatability.

**Photoautotrophic**: Green plants and photosynthetic bacteria are photoautotrophs. Photoautotrophic organisms are sometimes referred to as holophytic. Such organisms derive their energy for food synthesis from light and are capable of using carbon dioxide as their principal source of carbon.

**Photomixotrophically:** Deriving nourishment from both autotrophic and heterotrophic mechanisms- used especially of symbionts and partial parasites.

**Photosynthesis:** Formation of carbohydrate by the chlorophyll pigments using carbon dioxide and water in presence of light is termed as photosynthesis.

**Phytohormone:** The generic name for all classes of hormones produced by plants.

**Plant growth regulator:** It is a synthetic or natural chemical that exerts strong controlling effects on growth, development and metabolism at very low concentrations and usually at the site other than that of its synthesis.

**Plantlet:** A plant with a distinct root and shoot system developed through tissue culture propagation either by embryogenesis or organogenesis.

**Pollination:** Transfer pollen grains from anther to stigma.

**Primodium** (Primodia)**:** The earliest detectable stage of an organ.

**Proembryo:** Embryonic structure just preceding true embryo.

**Propagule:** Any plant part used for propagation.

**Protocorm:** Tuber-like structure formed when orchid seeds germinate or when meristem culture of orchid is done.

**Protocorm-like bodies:** Highly differentiated tissues (particularly of orchid) known as protocorm-like bodies (PLB), which can be induced through *in vitro* culture with the proper concentration and combination of plant growth regulators. Like field grown tubers, PLBs can develop into seedlings. The histological structure of PLBs are similar to those of for field grown tubers.

**Protoplast:** Plant cell without cell wall.

**RAPD (Random Amplified Polymorphic-DNA):** A PCR-based technique for detecting polymorphism at DNA level. In RAPD uses single short (usually 10-decamer primers) oligonucleotide primers for PCR.

**Recalcitrant seed:** The viability of this group of seeds drops drastically if their moisture content is reduced below 12-30%. Seeds of many forest and fruit trees and of several tropical crops like citrus, cocoa, coffee, rubber, oil palm, mango, jackfruit etc. belong to this group. Such seeds present considerable storage difficulties.

**Redifferentiation:** Differentiated plant cell or tissue has the ability to multiply into mass of undifferentiated cells, which is termed as dedifferentiation, and subsequently can undergo morphogenesis to regenerate a complete plant with all necessary organs of the mother plant is termed as redifferentiation.

**Reforestation:** Clearing of existing forest and replanting of forestry plants.

**Regeneration medium:** Synthetic plant tissue culture medium rich in cytokinin used for regeneration of plantlets from callus or direct organogenesis/ embryogenesis from explant.

**Regeneration:** Development and formation of new organs.

**Rhizogenesis:** It is a type of organogenesis by which only root initiation takes place in the callus mass.

**Seed:** Botanically seed is a mature ovule along with its food storage, either endosperm or cotyledon. In terms of seed science, seed can be described as any propagating material used for raising a crop.

**Self-incompatibility:** Pollen grain of same plant fails to fertilize the egg leading to no zygote formation.

**Self-pollination:** Pollen grains of flower pollinate the same flower.

**Shoot apex:** The apical meristem together with 1-3 young leaf primodia measuring 100-105 μm constitutes the shoot apex (see Fig. 2.5).

**Somaclonal variation:** Somaclonal variations have been defined as genetic and phenotypic variations among clonally propagated plants of a single donor clone and these are manifested as somatically or meiotically stable events. Larkin and Scowcroft (1981) for the first time used the word 'somaclonal variation' to describe these variations as displayed among the plants derived from tissue culture.

**Somatic embryo:** Somatic embryos are bipolar structures with both apical and basal meristematic regions, which are capable of forming shoot and root, respectively.

**Somatic embryogenesis:** It is the process of formation of a bipolar structure containing both shoot and root meristems either directly from the explant *de novo* origin from callus and cell culture induced from explant.

**Sterilization:** The procedure of elimination of organism from an object or from a micro-environment.

**Stock plant:** Plant from which explant is separated is known as stock plant.

**Stomata:** Stomata are a special type of pore opening found in the epidermis of leaves, stems, and other organs, that facilitates gas exchange. They are designed also to absorb water from sources such as rain while also removing excess water in the plant through transpiration.

**Subculture:** Transfer of the old-culture to a fresh medium.

**Suspension culture:** A cell suspension culture consists of cell aggregates dispersed and growing in moving liquid medium.

**Symbiotic:** The relationship between two living organisms of different species that live close together and depend on each other in various ways.

**Synthetic seed:** Encapsulated somatic-embryos or any other *in vitro* derived plant propagules in a nutrient gel which functionally mimic seeds and can develop into seedlings under sterile and/or natural conditions.

**Systemic:** Affecting or infecting whole of the body of a living organism.

**Thawing:** To change from a frozen solid to a liquid by gradual warming.

**Thermolabile:** A substance which disintegrates or is unstable upon heating.

**Tissue culture:** The *in vitro* culture of cell, tissue, organ, embryo, seed or protoplast on a nutrient medium under aseptic condition.

**Tissue:** A group of cells that perform a collective function.

**Totipotency:** It is the ability of a cell, tissue or organ to grow and develop a multi-cellular or multiorganed higher organism.

**Transgenic plants:** They are plants that have been genetically engineered, a breeding approach that uses recombinant DNA techniques to create plants with new characteristics. They are identified as a class of genetically modified organism (GMO).

**Transgenic:** An organism in which foreign gene(s) is incorporated into its genome. The transgene is inherited by offspring in Mendelian fashion.

**Vacuoles:** Vacuoles are storage bubbles found in cells. They are found in both animal and plant cells but are much larger in plant cells. Vacuoles might store food or any variety of nutrients a cell might need to survive. They can even store waste products so the rest of the cell is protected from contamination.

**Vermiculite:** It is a natural mineral that is often placed in potting soil to enhance a plants growth by boosting aeration while encouraging water drainage.

**Vitamins:** Naturally occurring organic substances, necessary in small amounts for the normal metabolism of organisms.

**Vitrification:** An undesirable condition of *in vitro* tissues characterized by succulence, brittleness and glassy appearance. The process of vitrification is a general term for a variety of physiological disorders that lead to a shoot tip and leaf necrosis. **OR** Vitrification in respect of cryopreservation refers to the physical process by which a highly concentrated cryoprotective solution supercools to very low temperature and finally solidifies into a metastable glass, without undergoing crystallization at a practical cooling rate. Thus, vitrification is an effective freez- avoidance mechanism.

**Weight:** Weight is a measure of the heaviness of an object. The force with which a body is attracted to Earth or another celestial body, equal to the product of the object's mass and the acceleration of gravity.

**Zygotic embryo:** A young sporophytic plant, as yet retained in the gametophyte in the seed.

# 1
# Introduction

## Introduction

Progress in plant biotechnological research has opened many avenues for basic and applied research in the field of crop plants. Plant tissue culture is an important component of biotechnology that involves in the improvement of crops. Besides this, plant tissue culture provides a good system for many basic studies in plant breeding, plant physiology, genetics and cell biology. Further, manipulation of cells through the modern technique of genetic engineering may lead to an introduction of new gene(s) to a host plant and the techniques of genetic engineering rely on plant tissue culture techniques. The regeneration of plants through plant tissue culture and their subsequent acclimatization and delivery to the field possesses many problems to make tissue culture technology a viable alternative proposition. The successful development of encapsulation technique for preparation of plant propagules in a nutrient gel has initiated a new line of research on ***artificial seeds***. Application of this technology has been well recognized in several agronomically important crops and forest trees. Artificial seeds make a promising technique for propagation of transgenic plants, non-seeds producing plants, polyploids with elite traits and plant lines with problems in seed propagation.

## 1.2. Landmark of Contribution of Plant Tissue Culture

The idea of artificial seed was first conceived by Toshio Murashige which was subsequently developed by several investigators (Bapat, 1993; Datta and Potrykus, 1989), and besides rapid and mass propagation of plants, the artificial seed technology has added new dimensions not only to handling and transplantations but also for conservation of endangered and desirable genotypes. During seventies of twentieth century, he tried to explain the concept of artificial seed preparation and its possible use in clonal propagation of plants. He has presented many papers in seminars and symposium on tissue culture propagation where he concluded with the concept of artificial seed. His presentation on artificial seeds at the 'Symposium on Tissue Culture for Horticultural Purposes in Ghent, Belgium', September 6-9, 1977 opened doors for further research on artificial seeds. His comment on the artificial seeds on the Symposium is remarkable, "..... *the cloning method must be extremely rapid, capable of generating several million plants daily and competitive economically with the seed method*" (Murashige, 1977). Subsequently many research workers emphasized the need for research on artificial seeds and outlined its impacts on mass propagation in true breeding of crop plants (Brar *et al*., 1994). The major developments or events in plant tissue culture are being listed in Table 1.1.

**Table 1.1.** Landmark of contribution of plant tissue culture

| Year | Event |
|---|---|
| 1883 | Totipotency theory (Schwann and Scheilden) – cells are autonomic, and in principle, are capable of regenerating to give a complete plant |
| 1892 | Plants synthesize organ forming substances which are polarly distributed (Sachs) |
| 1902 | German botanist Gottlieb Harberlandt developed the concept of in vitro cell culture |
| 1909 | Fusion of plant protoplasts, although the products failed to survive (Kuster) |
| 1904 | Hanning initiated a new line of investigation involving the culture of embryogenic tissue |
| 1922 | Small excised root tips of pea and maize cultivated by Kotte.<br>At the same time USA scientist, Robbins also cultured root tips of corn on artificial medium |
| 1925 | Laibach (1925, 1929) demonstrated the practical applications of zygotic embryo culture in the field of plant breeding |
| 1929 | Embryo culture of Linum to avoid cross incompatibility (Laibach) |
| 1934 | White (1934, 1937) successfully established the plant root culture<br>Gauthreret cultured cambium cells of some tree species |
| 1936 | Embryo culture of various gymnosperms (LaRue) |
| 1939 | Gautheret successfully established the continuously growing callus culture |
| 1940 | *In vitro* culture of cambial tissues of Ulmus to study adventitious shoot formation (Gautheret) |
| 1941 | Coconut milk used by Van Overbeek*et al.* for the first time in *Datura* tissue culture and a cell division factor detected in it |
| 1942 | Gautheret observed the secondary metabolites in plant cell cultures |
| 1944 | First in vitro cultures of tobacco used to study adventitious shoot formation (Skoog) |
| 1945 | Cultivation of excised stem tips of Asparagus in vitro (Loo) |
| 1946 | First whole Lupinus and Tropaeolum plants from shoot tips (Ball) |
| 1948 | Ormation of adventitious shoots and roots of tobacco determined by the ratio of auxin/adenin(Skoog and Tsui) |

1950 Organs regenerated from callus tissue of Sequoia sempervirens (Ball)

1951 Nistch successfully cultured excised ovule *in vitro*

1952 Morel and Martin developed protocol for meristem culture to obtain virus-free plants in Dahlias

1953 Production of haploid callus of the gymnosperm *Ginkobiloba* from pollen (Tulecke)

1954 Monitoring of changes in karyology and in chromosome behavior of endosperm cultures of maize (Strauss)

1955 Miller *et al.* discovered kinetin, a cell division hormone

1956 Realization of growth cultures in multi-litre suspension systems to produce secondary products by Tulecke and Nickell (Staba, 1985)

1957 Skoog and Miller proposed the concept of hormonal control of organ formation

1958 Regeneration of somatic embryos in vitro from the nucellus of Citrus ovules (Maheshwari and Rangaswamy). Regeneration of pro-embryos from callus clumps and cell suspensions of Daucuscarota (Reinert, Steward)

1959 Regeneration of somatic embryos from callus clumps and cell suspension of *Daucuscarota* (Reinert)
Braun demonstrated the recovery of crown gall tumor cell *in vitro*

1960 Cocking isolated protoplast using method of enzymatic degradation of cell wall
Morel produced virus free orchid plants through meristem culture

1962 Murashige and Skoog developed synthetic nutrient medium for plant tissue culture
The initial intra-ovarian pollination led to the subsequent development of test-tube fertilization (Kanta*et al.*)

1964 Guha and Maheswari (1964, 1966) regenerated haploid plants from microspore culture of *Datura anoxia*

1965 Differentiation of tobacco plants from a single isolated cell (Vasil and Hildebrandt)

1967 Flower induction in Lunariaannua by vernalisation in vitro (Pierik).Haploid plants obtained from pollen grains of tobacco (Bourgin and Nitsch)

1969 Karyological analysis of plants regenerated from callus cultures of tobacco (acristan and Melchers). First successful isolation of protoplasts from a suspension culture of Hapopappusgracilis (Eriksson and Jonassen)

1970 Fusion of protoplast was done by Power and his coworkers

1971 Takebe and his coworkers for the first time regenerated plants from protoplasts

1972 First report of regeneration of somatic hybrid plants through protoplast fusion in two species of *Nicotiana* (Carlson and his coworkers)

1973 Cytokinin found capable of breaking dormancy in excised capitulum explants of Gerbera (Murashige et al.)

1974 Induction of axillary branching by cytokinin in excised Gerbera shoot tips (Murashige et al.). Regeneration of haploid Petunia hybrida plants from protoplasts (Binding). Fusion of haploid protoplasts found possible which gave rise to hybrids (Melchers and Labib). Biotransformation in plant tissue cultures (Reinhard). Discovery that the Ti-plasmid was the tumour inducing principle of Agrobacterium (Zaenen et al.; Larebeke et al.)

| | |
|---|---|
| 1975 | Positive selection of maize callus cultures resistant to Helmintho sporiummaydis (Gengenbach en Green) |
| 1976 | Seibert reported shoot initiation from cryopreserved shoot apices of carnation |
| 1978 | Somatic hybridization between distantly related species like tomato and potato, resulting pomato (Melchers and his coworkers) |
| 1981 | Introduction of the term somaclonal variation (Larkin and Scowcroft) |
| 1982 | Synthetic seed was produced first time by Kitto and Janick (1982, 1985) involving carrot somatic embryos |
| 1983 | Intergeneric cytoplasmic hybridization in radish and rape (Pelletier et al.) |
| 1984 | Transformation of plant cells with plasmid DNA (Paszkowski et al.)<br>Development of the genetic fingerprinting technique for identifying individuals by analyzing Polymorphism at DNA sequence level (Alec Jeffreys) |
| 1985 | Infection and transformation of leaf discs with Agrobacterium tumefaciens and the regeneration of transformed plants (Horsch et al.) |
| 1986 | Release of first herbicide resistant GM tobacco<br>TMV virus-resistant tobacco and tomato transgenic plants developed using cDNA of coat protein gene of TMV (Powell-Abel et al.) |
| 1987 | Development of biolistic gene transfer method for plant transformation(Sanford et al.; Klein et al.)<br>Isolation of Bt gene for bacterium (Bacillus thuringiensis) (Barton et al.) |
| 1990 | First successful field trial of herbicide resistant GM cotton |
| 1993 | Janick and his coworkers have reported that coating a mixture of carrot somatic embryos and callus in polyoxyethelene glycol produced desiccated artificial seeds |
| 1994 | 'FlavrSavr' tomato becomes the first GM food approved for sale |
| 2001 | Development of genetically engineered 'Golden Rice' |

## 1.3. Definition and Brief History

Botanically ***seed*** is a matured ovule along with its food storage, either ***endosperm*** or ***cotyledon***. Seeds are desiccation tolerant, durable and quiescent due to a protective seed coat. By means of seed, plants are able to transmit their genetic constitution in generations and therefore seeds are the most appropriate means of propagation, storage and dispersal (Bewley and Black, 1985). The essential part of seed is the embryo contained within the integuments, but it may be used less critically to describe planting materials. In terms of seed science, seed can be described as any propagating material used for raising a crop. Whereas, ***artificial seed*** could be defined as artificially encapsulated somatic embryos, shoot buds, cell aggregates or any other tissue that can be used for sowing as a seed and that possesses the ability to be converted into a plant under *in vitro* or *ex-vitro* conditions, and that retains this potential even after storage (Capuano *et al.,* 1998).

Murashige (1977) was the first to produce an official definition of artificial seed (synthetic seed or synseed)- 'An encapsulated single somatic embryo, i.e., a clonal product that could be handled and used as a real seed for transport, storage and sowing, and that, therefore, would eventually grow, either *in vivo* or *ex vitro*, into a plantlet'. An artificial seed was later defined by Gray *et al.* (1991) as 'A somatic embryo that is engineered for the practical use in commercial plant production'. According to Bekheet (2006), 'An artificial seed is often described as a novel analogue to true seed consisting of a somatic embryo surrounded by an artificial coat which is at most equivalent to an immature zygotic embryo, possibly at post-heart stage or early cotyledonary stage'. The definition of artificial seeds was then extended to be artificially coated somatic embryos or other vegetative parts such as shoot buds, cell aggregates, auxiliary buds, or any other micropropagules, provided that they have the capacity to be sown as a seed and converted into a plant under in vitro or ex vitro conditions. They should also be able to keep this ability for an extended period of storage (Bapat *et al.*, 1987; Ara *et al.*, 2000; Daud *et al.*, 2008).

Somatic embryos are the main source of artificial seed production; however, other micropropagules like shoot buds, shoot tips, organogenic or embryogeniccalli, protocorms, protocorm-like bodies etc. have also been employed in the production of artificial seeds. Production of artificial seeds endowed with high germination rate under *in vitro* and *in vivo* conditions, bears immense potential as an alternative of true seeds. It is an emerging area with great potential for large-scale production of ***propagules*** at lower cost with ease in handling and transport of crop plants (Ara *et al,* 1999; Redenbaugh *et al.,* 1993; Roy and Mandal, 2008).

Implementation of artificial seed technology requires manipulation of *in vitro* culture systems for large-scale production of viable materials that are able to convert into plants, for encapsulation. For most of the crop plants, techniques have been standardized for large-scale somatic embryo production from different explants. High frequency somatic embryogenesis, synchronous embryo development and maximum conversion of embryos into plantlets have been very critical for realizing the production of artificial seeds (Senaratna, 1992). Naked micropropagules (somatic embryos, pro-embryos, embryo-like structure, shoot buds, protocorms, protocorm-like bodies etc.) are sensitive to desiccation and/or pathogens when exposed to natural environment. Thus alginate gel on encapsulation provides protection to the naked somatic embryos and other propagating materials and also facilitates its handling (Redenbaugh and Walker, 1990). Encapsulation of embryo may control water uptake, release of nutrients and provide mechanical protection required for field planting.

Production of artificial seed has opened a new avenue in rapid and true-to-type multiplication of elite varieties of crops. The technology can also help in germplasm storage and transportation of elite genotypes. A strong potential exists for propagation of high yielding, individual hybrids through somatic embryogenesis and artificial seeds (Brar *et al.*,1994). For potential application of seed encapsulation, technology has been demonstrated for many crop plants (Bapat and Rao, 1988, Padmaja *et al.,* 1995; Onay *et al,* 1996; Shigeta and Sato, 1994; Suprasanna *et al.,* 1996).

Development of artificial seed production technology is currently considered as an effective and alternate method of propagation in several commercially agronomic and horticultural crops. It has been suggested as a powerful tool for mass propagation of elite plant species with high commercial value (Saiprasd, 2008). It is the most effective techniques for the propagation of plant species that have problems in seed propagation and plants that produce non-viable seeds (Daud *et al*., 2008). Currently the artificial seed production technology have progressed substantially, the most advanced being in seeding under *ex vitro* obtaining high percentage of conversion to plants (Nieves *et al*., 2003a). However the prime objective of this technology to produce an analogue to natural seed is yet to be achieved. In this endeavour, efforts have been made to aggregate the research findings on artificial seed technology and these information have been presented in systematic manner in different chapters.

## 1.4. Aim and Scope of Artificial Seeds

Synthetic seed technology is the most significant applications of plant tissue culture, have great potential for large scale production of different types of plants at low cost as an alternative to true seeds. From the biotechnological point of view, there are numerous advantages of artificial seeds such as ease of handling, low production cost, ease of exchange of plant materials, genetic uniformity of plantlets, direct delivery to the soil, shorten the breeding cycle and reduction of the storage space. Apart from it, there are several other benefits such as large scale propagation method, rapid multiplication of plants used in advanced procedures of cryopreservation for the long term preservation of plant germplasm etc. Basic aims of this technology are to convert different micropropagules, mainly somatic embryo into artificial or synthetic seeds which can be grown in the field or greenhouse when required as well as better clonal parts could be propagated similar to seeds, preservation of rare plant species and more consistent and synchronised harvesting of important agricultural crops would become a reality among many other possibilities. Among crop plants, this technique can be used for plants which do not multiply by true seeds or the seeds if available, the prices are not affordable.

Micropropagation through artificial seeds may be commercially exploited on a large scale and this technology would be feasible and even competitive economically with the traditional seed propagation.The technology provides methods for preparation of seed analogues called synthetic seeds or artificial seeds from the micropropagules like somatic embryos, axillary shoot buds, apical shoot tips, embryogeniccalli as well as protocorm or protocorm like bodies (Ara *et al.*, 2000; Danso and Ford-Lloyd, 2003; Rai *et al.*, 2008; Roy and Mandal 2008; Ahmad and Anis, 2010; Rihan *et al.*, 2011; Sharma and Shahzad, 2012). Implementation of synthetic seed technology requires manipulation of *in vitro* culture systems for large-scale production of viable materials that are able to convert into plants, for encapsulation. Through these systems a large number of somatic embryos or shoot buds are produced which are used as efficient planting material as they are potent structures for plant regeneration either after having minor treatment or without any treatment with growth regulator(s).

Because of the naked micropropagules that are sensitive to desiccation and/or pathogens when exposed to natural environment, it is envisaged that for large scale mechanical planting and to improve the success of plant (*in vitro* derived) delivery to the field or greenhouse, the somatic embryos or even the other micropropagules useful in synthetic seed production would necessarily require some protective coatings. The primary goal of synthetic seed research as so far concerned is to produce somatic embryos that resemble more closely the seed embryos in storage and handling characteristics so that they can be utilized as a unit for clonal plant propagation and germplasm conservation.

Either hydrated calcium alginate-based or desiccated polyoxyethylene glycol-based artificial seeds might be used, but it is likely that some degree of drying before cryopreservation would be beneficial. In some crop species, seed propagation is not successful. This is mainly due to heterozygosity of seeds, minute seed structure etc. For example, in case of orchids, presence of reduced endosperm, requirement of mycorrhizal fungus association with seed for germination (orchid) and also some seedless variety of crop species like grapes, watermelon etc. Some of these species can be propagated by vegetative means, however *invivo* vegetative technique are time consuming and expensive. Therefore, the development of artificial seed production technology is currently considered as an effective and alternative method of production in several commercial important agronomic and inter-culture crops. It is a powerful tool for mass propagation of superior plant species with high commercial value.

The synthetic seed is currently considered as the most effective and efficient alternate technique for propagation of commercially important plant species that had problems in seed propagation and plants that produced non-viable seeds or without seeds. Zygotic embryos are formed from the sexual recombination of male and female gametes. Thus, cuttings or other vegetative means are used for propagation and these methods are rarely present for convenient storage. Artificial seeds could be a good tool to propagate such types of plants and to store their propagules for a considerable period of time.

Artificial seed production is considered to be an authentic technique for the proliferation of plant species which are not able to produce seed, such as seedless grapes and seedless watermelon. Artificial seeds can be employed for production of polyploids with elite traits, avoiding the genetic recombination when these plants are propagated using conventional plant breeding systems, thus saving on time and costs. Artificial seeds can be also used in the propagation of male or female sterile plants for hybrid seed production.

Artificial seed production through the use of somatic embryos is an important technique for transgenic plants, where a single gene can be placed in a somatic cell and this gene will be located in all the plants which are produced from this cell. Therefore, artificial seeds could also be a resourceful technology used for reproduction of transgenic plants. The encapsulation technology can be considered as a promising and a challenging approach that can be used for the exchange of plant materials in tissue culture laboratories and also to achieve germplasm conservation and the propagules that are derived from *in vitro* or by micropropagation that can be used directly in nurseries or in a field. Moreover, artificial seeds, which are produced using tissue culture techniques that are aseptic, are free of pathogens, giving great advantages to these materials for transport across frontiers and for avoiding the spread of plant diseases.

Artificial seeds are also valuable in terms of their role in providing protective coating, increasing the level of micropropagule success in the field. These micropropagules need a protective coating to increase successful establishment under the field conditions because of the sensitivity of uncovered micropropagules to drought and pathogens under natural environmental conditions. Artificial seed production is also a useful technique as a clonal propagation system in terms of preservation of the genetic uniformity of plants, straight delivery to the field, low cost, and fast reproduction of plants. Moreover, the use of this technique economizes upon the space, medium, and time requested by the traditional tissue culture methods. Apart from its scope, artificial seed production has great advantages in comparison with traditional tissue culture methods.

Artificial seeds (encapsulated somatic embryos, in specific) can open new vistas for land restoration and the rehabilitation of wild lands (rangelands, grasslands, forests, abandoned mine lands, etc.) affected by overgrazing or climate change. The main objective of artificial seed research is the production of an artificial seed structure that stimulates the conventional seed in its characteristics (such as handling, storage, viability, and germination level).

Cryopreserved artificial seeds may also be used for germplasm preservation particularly in recalcitrant species. For example: Mango, coconut etc. as these species will not undergo dessication. It is a potential tool in the genetic engineering of plants because a single gene can be inserted into somatic cell and propagated by synthetic seeds.

Multiplication of superior plants selected in plant breeding programme through somatic embryos avoids a genetic recombination and therefore saves time and resources. They are free of pathogen. Thus the transport of pathogen free propagules across the international borders is possible avoiding bulk transportation of plants, quarantine and spread of diseases. This technique has great advantages such as: a cost-effective delivery system, minimization of the cost of plantlets, simple methodology with high potential for mass production, a promising technique for the direct use of artificial seedlings in vivo, and a high storage capacity.

## References

Ahmad N, Anis M. 2010. Direct plant regeneration from encapsulated nodal segments of *Vitexnegundo*. Biol Plant. 54: 748-752.

Ara H, Jaiswal U, Jaiswal V. 2000. Synthetic seed: Prospects and limitation. Curr Sci. 78: 1438-1444.

Ara H, Jaiswal U, Jaiswal VS. 1999. Germination and plantlet regeneration from encapsulated somatic embryos of mango (*Mangifera* indica L.).Plant Cell Rep.19: 166-170.

Bapat VA, Mhatre M, Rao PS. 1987. Propagation of *Morusindica* L. (Mulberry) by encapsulated shoot buds. Plant Cell Rep. 6: 393-395.

Bapat VA, Rao PS. 1988. Sandalwood plantlets from synthetic seeds.Plant Cell Rep. 7: 434-436.

Bapat VA. 1993. Studies on synthetic seeds of sandalwood (*Santalum album* L.) and mulberry (*Morusindica* L.). In: Redenbaugh, K. ed. Synseeds: applications of synthetic seeds to crop improvement. Boca Raton, CA, United States, CRC Press Inc. pp. 381-407.

Bekheet SA. 2006. A synthetic seed method through encapsulation of *in vitro* proliferated bulblets of garlic (*Allium sativum* L.). Arab J Biotech.9: 415-426.

Bewley JD, Black M. 1985. Seeds: Physiology of Development and germination, Plenum Press, New York. pp. 367.

Brar DS, Fujimura T, McCouch S, Zapate FJ. 1994. Application of biotechnology in hybrid rice. In: Hybrid Rice Technology: New Development and Future Prospects, Virmani SS (ed.), International Rice Research Institute, Manila, Philippines. pp. 51-62.

Capuano G, Piccioni E, Standardi A. 1998. Effect of the different treatments on conversion of M26 rootstocks synthetic seeds obtained from encapsulated apical and axillary micropropagated buds. J. HorticSciBiotechnol.73: 299-305.

Danso KE, Ford-Lloyd BV. 2003. Encapsulation of nodal cuttings and shoot tips for storage and exchange of cassava germplasm. Plant Cell Rep. 21: 718-725.

Datta SK, Potrykus I. 1989. Artificial Seeds in Barley: Encapsulation of Microspore-Derived Embryos. TheorAppl Genet. 77: 820-824.

Daud M, Taha MZ, Hasbullah AZ. 2008. Artificial seed production from encapsulated micro shoots of *Sainpauliaionantha* Wendl. (African Violet). J Appl Sci. 8: 4662-4667.

Gray DJ, Purohit A, Triglano RN. 1991. Somatic embryogenesis and development of synthetic seed technology. Crit Rev Plant Sci. 10: 33-61.

Murshige T. 1977. Plant cell and organ cultures as horticultural practice. Acta Hortic. 78: 17-27.

Nieves N, Zambrano Y, Tapia R, Cid M, Pina D, Castillo R. 2003a. Field performance of sugarcane plants obtained from artificial seed in Cuba. Sugarcane International, November, 2003. pp. 23-25.

Onay A, Jeffree CE, Yeoman MM. 1996. Plant regeneration from encapsulated embryoids and an embryogenic mass of pistachio (*Pistacivera* L.). Plant Cell Rep. 15: 723-726.

Padmaja G, Reddy LR, Reddy GM. 1995. Plant regeneration from synthetic seeds of groundnut (*Arachishypogaea* L.). Indian J Expt Biol. 33: 967-971.

Rai MK, Jaiswal VS, Jaiswal U. 2008. Encapsulation of shoot tips of guava (Psidiumguajava L.) for short-term storage and germplasm exchange. Sci Hortic. 118: 33-38.

Redenbaugh K, Fujii JA, Slade D. 1993. Hydrated coatings for synthetic seeds. In: Synseeds: application of synthetic seeds to crop improvement, Redenbaugh K (ed.). CRC, Boca Raton. pp. 35-46.

Redenbaugh K, Walker K. 1990. Role of artificial seeds in alfalfa breeding. In: Plant tissue culture: application and limitations. Development in crop science, Bhojwani SS (ed.). 19-Elsevier, Amsterdam, pp. 102-135.

Rihan HZ, Al-Issawi M, Burchett S, Fuller MP. 2011. Encapsulation of cauliflower (Brassica oleraceavar botrytis) microshoots as artificial seeds and their conversion and growth in commercial substrates. Plant Cell Tissue Organ Cult. 107: 243-250.

Roy B, Mandal AB. 2008. Development of synthesis seed involving androgenic and pro-embryos in elite *indica* rice. Indian J Biotechnol.7: 515-519.

Saiprasad GVS. 2008. Artificial seeds and their application. Resonance. 6: 39-47.

Senaratna T. 1992. Artificial seeds.Biotech Advance. 10(3): 379-392.

Sharma S, Shahzad A. 2012. Encapsulation technology for short-term storage and conservation of a woody climber, *Decalepishamiltonii* Wight and Arn Plant Cell Tissue Organ Cult. 111: 191-198.

Shigeta J, Sato K. 1994. Plant regeneration and encapsulation of somatic embryos of horse radish.Plant Science.102: 109.

Suprasanna P, Ganapathi TR, Rao PS. 1996. Artificial seed in rice (*Oryza sativa* L.): Encapsulation of somatic embryos from mature embryo callus culture. Asia Pacific J MolBiotechnol. 4(2): 90-93.

# 2

# Fundamentals of Plant Tissue Culture

## 2.1. LABORATORY SETUP

The setting up of a tissue culture laboratory needs proper planning. It depends on the nature of research works to be under taken and/or commercialization of tissue culture technology and the availability of space. The ideal tissue culture laboratory should have the following components:

1. Store facility for chemicals
2. Media preparation room
3. Autoclave room
4. Storage facilities for autoclaved laboratory-ware and media
5. Washing facilit
6. Culture room
7. Inoculation room
8. Instrumentation room
9. Hardening facility

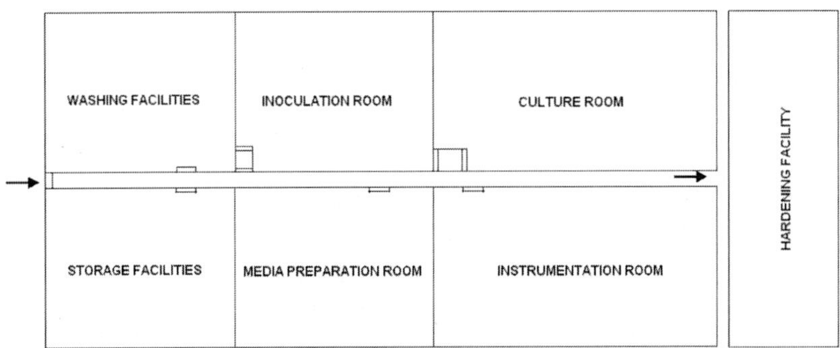


**Fig. 1.1.** Layout of an ideal Tissue Culture Laboratory

### 2.1.1. Storage Facility

For media preparation, a large number of chemicals are required, which includes inorganic salts, plant growth regulators, organic supplements, liquid chemicals etc. Separate store place is essential for liquid chemicals, because they are mostly in glass containers. Accident may take place on breakage of these glass bottles. The chemicals should be arranged in alphabetical order, so that it can be easily found out when it is required. Replace the bottle/containers after use at the appropriate place.

### 2.1.2. Labware Cleaning Facility

Washing area should be large enough to accommodate a big size sink. Provisions of hot and cold running water with proper drainage facilities are essential. It should also have space for demineralized, distilled and double ***distilled water*** facilities and an oven for drying of cleaned and washed glassware. The other accessories of washing room are plastic buckets and tubs for soaking laboratory-ware, acid or detergent bath, pipette washers, brushes for culture tube, conical flasks and beaker washing. For storage of washed and dried laboratory-ware, washing room may be provided with dustproof storage cabinets. The method and precaution for cleaning of glass- and plastic-wares are different, thus the cleaning methods have been described separately.

#### *2.1.2.1. Glassware Cleaning and Washing*

Varieties of glassware are being used in tissue culture laboratory, viz. beakers, conical flask, measuring cylinder, culture bottles, culture tubes, reagent bottles, etc. The glassware should be of good quality, heat resistant and less prone to breakage. Generally glassware are soaked in concentrated sulphuric acid for 4 hours subsequently through-wash in tap water followed by distilled water. As acid is corrosive in nature, high precaution is needed, such as wearing of hand gloves, plastic/cloth apron etc. To avoid these difficulties, detergent wash is preferred in place of acid wash. Glassware is soaked in detergent solution for 16-20 hours. The glassware removed from the culture room containing used agar media and/or contaminated by fungus or bacteria should be boiled/ autoclaved before soaking in detergent solution, then rinsed in tap water followed by rinsing in distilled water.

The cleaned glassware is dried in convection-drying oven at 160-180 °C for 2-4 hours. The cleaned and dried glass-wares may be directly taken for use or store in dust proof cabinet.

#### *2.1.2.2. Plastic-wares Cleaning*

To reduce the risk of glassware breakage, use of plastic-ware increased drastically. The commonly used plastic-wares are beaker, measuring cylinder, culture bottle, Petri-plates etc. Used plastic-wares are soaked in non-corrosive detergent and washed in tap water followed by rinsing in distilled water.

Few precautions are to be taken before and/or during laboratory-ware claiming:

1. The measuring cylinder used to prepare stock/medium should be rinsed in tap water after use.
2. Immediately after dispensing medium from beaker/conical flask to culture tubes/ Petri-dishes, the beaker should be washed in tap water to avoid sticking of agar to the walls of laboratory-ware.
3. Contaminated laboratory-ware should be autoclaved before bringing to washroom.
4. Avoid rinsing of plastic-wares with organic solvents, such as acetone, chloroform or methyl chloride.
5. Remove labels and marking ink before soaking in detergent solution.

#### *2.1.2.3. Hot Air Oven*

Hot air oven is mainly used for drying of glass-wares after washing. It can also be used for dry heat sterilization of cleaned glassware. It is thermostatically controlled, which ranges from 5 °C to 250 °C. The detail of dry sterilization has been discussed in section 2.2.1.2 in this chapter.

### 2.1.3. Media Preparation Room

The most important technique for media preparation is the bench space for chemicals, laboratory-ware, culture vessels, and culture tube stands etc. It should have space to accommodate some essential equipment, such as electronic balance of high resolution, pH meter, magnetic stirrer with hot plate, vortex, refrigerator, microwave oven, water bath, and gas (LPG) burner/stove. When the work-load is not at commercial scale, a domestic pressure cooker of 22 liter can be used in place of vertical or horizontal autoclaves to autoclave media and other laboratory-ware. A dust proof storage cabinet may be kept in this room to store autoclave media and laboratory-ware.

Activities performed in the media preparation room:

- Preparation of stock solutions
- Mixing of stock solutions for preparation of media
- Melting and dispensing of media into culture tubes
- Autoclaving of media of non-commercial scale

### 2.1.4. Autoclave Room

During autoclaving, a high pressure is developed inside the autoclave cylinder. The pressure inside the autoclave is maintained by using a knob. If it is unattained to control the pressure, inside pressure rises high and high, this may lead to further bursting the autoclave cylinder. To avoid causality due to this type of accident, there should be a separate autoclave room.

Autoclave or pressure cooker is used to sterilize culture media in glass-vessels (culture tubes, culture bottles, conical flasks etc.), water, small instruments or equipments (such as, different types of filters etc.) and materials to be used during ***inoculation*** of explants (such as, tissue paper, filter papers, aluminum foils, forceps, scalpel, spatula, needles, glass rod, empty glassware, autoclave-able plastic ware, etc.). The required temperature and duration of autoclaving the medium have been discussed in section 2.2.1.3.

### 2.1.5. Inoculation Room

Inoculation room is the most important component of an ideal tissue culture laboratory. Explants are sterilized and transferred/inoculated on culture medium under laminar air flow cabinet in this room. It is a dust-free room equipped with air-conditioner, laminar air flow cabinet, and vacuum pressure machine for filter sterilization. Aseptic condition is the main motto of this room. The entry has double door with air-bath facility to reduce the entry of contaminants in tissue culture room. At least two UV lights on the both sides of laminar air flow cabinet is essential.

Filter sterilization unit with or without vacuum pump kept in inoculation room for filter sterilization of thermo-sensitive chemicals, such as antibiotics and hormones.

### 2.1.6. Culture Room

*In vitro* plant growth and/or callus development is greatly influenced by the microenvironment in the culture vessels as well as different atmospheric parameters of the culture room. The most important atmospheric parameters to be considered in this room are light/ dark period, temperature and humidity. In an ideal culture room, the temperature is maintained around 25 ± 2 °C. Nowadays air conditioners are available with temperature sensor to maintain a particular (fixed) temperature in the culture room. Light quality, intensity and duration are to be maintained in this room. Timer is being used in culture room to maintain 16/8 hours light and dark cycle. Most of the culture room is lighted at 1000 lux with some going up to 10000 lux with wide spectrum or cool white fluorescent lights.

Cultures are kept on racks inside the culture room. Very sophisticated racks are available with timer to control light duration. Culture vessels (culture tubes, culture bottles, culture plate) can be placed directly on the selves/racks on trays (especially for culture plates) of a suitable size. All the culture should be labeled appropriately and kept on the racks on a specified place. Continuous dark is required for callus induction; therefore bottom two selves of a racke at corner of culture room should be covered with thick black cloth/curtain. Under light, chlorophyll synthesis takes place leading to start the process of photosynthesis by the cells with chlorophyll. This inhibits the morphogenetic ability of callus.

Continuous power supply is very essential for culture room. Therefore, in case of power failure, an alternative arrangement of generator is required.

### 2.1.7. Instruments Room

A number of instruments are being used to run a modern tissue culture laboratory. Thus, a separate instrumentation room is very much essential. The instruments are arranged based on the uses and their functions. Instruments usually required for tissue culture laboratory are as follow:

#### *2.1.7.1. Microscopes*

a) Stereomicroscope: It is used to study the surface of small objects. The microscope should have the camera attachment to take the photograph of the specimen. In tissue culture laboratory it is used for the following purpose:

- To count the number of embryo on embryogenic callus
- To study the surface nature of callus

- To study in detail the external progress of organogenesis and embryogenesis.

b) Compound microscope: Compound microscope is used to study the anatomy and cytology of callus and regenerated ***plantlets***. Anther culture and ovule culture may produce haploids and other higher ploidy in some cases. Thus, compound microscope is used to study the cytology (chromosome number) of the callus or regenerated plantlets. Compound microscope is also used to find out the appropriate stage of microspore for anther culture or microspore culture.

### *2.1.7.2. Electronic Balance*

High quality electronic balance is required to measure the small quantity of chemical compounds. The electronic balance is user friendly. It can be used more quickly and efficiently for sensitive quantity. It should be sensitive to even milligram quantity of chemical and the upper limit may be of 100 gram.

### *2.1.7.3. pH Meter*

The pH meter is used to measure the pH of various tissue culture media and stock solutions. The pH of the tissue culture medium usually varies from 5.0-5.8. Sometimes lower pH is also required for special cases, for example, in vitro screening for Al-toxicity tolerant genotypes, the required pH of the medium is < 5.0.

### *2.1.7.4. Magnetic Stirrer with Hot Plate*

Some chemicals do not readily dissolve in water; need continuous stirring for few minutes to even hours to dissolve it. Some other chemicals may also need continuous stirring as well as heating to dissolve in water. For example, to dissolve sodium alginate with tissue culture medium, the solution needs to be heated along with continuous stirring on hot plate. Thus, magnetic stirrer with hot plate is an inseparable instrument of tissue culture laboratory.

### *2.1.7.5. Distillation Set*

For preparation of media and stocks for tissue cultural works, it needs a continuous supply of good quality water. The final wash of glass-wares and plastic-wares is done with distilled water. A distillation unit or deionizer is essential in tissue culture laboratory.

*2.1.7.6. Vacuum Cleaner*

Inoculation and culture rooms must be dust free. To maintain cleanliness in these rooms vacuum cleaner is to be used regularly. Using vacuum cleaner in all corners, culture racks and floors can be kept dust free.

*2.1.7.7. Refrigerator/freezer*

Refrigerator is used to keep thermo-sensitive chemicals, such as hormones, enzymes, readymade kits etc. and stock solutions.

*2.1.7.8. Shaker*

Gyratory shaker with platform and clips for different size is required for cell suspension culture and bacteria culture for transformation works. The shaker should be placed in an area with controlled light and temperature. Nowadays closed chamber shaker is also available in which temperature and light intensity can be adjusted as per-requirement.

*2.1.7.9. Glass Bead Sterilizer*

The glass bead sterilizer uses heated glass beads for fast sterilization of small laboratories equipments. The instrument has a compact design effectively sterilizes small solid metal and glassware items within 10 seconds. When glass beads are heated to 300 °C, it destroys bacteria and spores. The instrument is used for forceps, scissors, scalpels, needles, ring vaccination and inoculation needles. It is used in research laboratories.

*2.1.7.10. Other Working Tools*

Other working tools include forceps, scalpel, spatula, culture tubes racks for stab and slant culture, teflon sieves of various pore sizes (40-250 μ mesh screens) to separate cell clumps of different dimensions from suspension culture, alcohol burner or gas (LPG) burner, parafilm for wrapping *Petri*-plates, plastic trays for transfer of cultures from one room to another, timer to regulate light period of culture room etc.

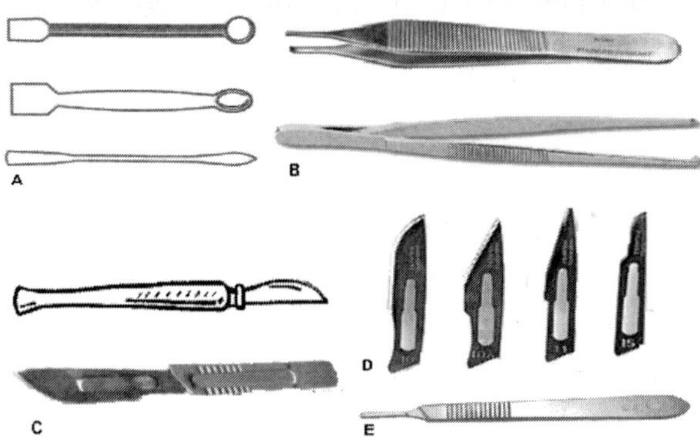


**Fig. 2.2.** Different types of tools used in plant tissue culture.
A) Spatulas, B) Forceps, C) Scalpels with blade, D) Blade of scalpel and E) Handle of scalpel.

#### *2.1.7.11. Other Instruments*

Which are often required for tissue cultural laboratory for some specific purposes are hemocytometer to determine cell counts, air conditioner to regulate temperature of inoculation and culture room, ***PCR*** machine, cooling centrifuge, table-top centrifuge, gel ***electrophoresis***, gel documentation system, etc.

## 2.2. MAINTENANCE OF ASEPTIC CONDITIONS AND STERILIZATION TECHNIQUES

Microbes are ubiquitous in nature. Bacteria and fungus are the common contaminants in tissue culture process. Their growth in the tissue culture medium is faster than the growth of explants. Usually these microorganisms are ***parasitic*** in nature to the explants and they directly take nutrients from the medium provided for tissue culture materials. Sometimes they produce chemical(s) which may become toxic to tissue culture materials. Therefore, maintenance of aseptic conditions in inoculation and culture rooms is very much essential.

Besides culture room and inoculation room, all the laboratory-wares and culture media used for tissue culture should be pathogen free. Different laboratory-ware, media and some specific chemicals which cannot be autoclaved are sterilized in different ways and are being elaborated below.

### 2.2.1. Methods of Aseptic Manipulation and Sterilization

#### *2.2.1.1. Maintenance of Aseptic Condition in Inoculation Room*

Transfer room or inoculation room is the most important component of ideal tissue culture laboratory. It is a dust-free room equipped with air-conditioner, laminar air flow cabinet, and vacuum pressure machine for filter sterilization. Aseptic condition is the main motto of this room. The entry has double door with air bath facility to reduce the entry of contaminants. Open the first door and enter into the space in between the two doors, take air bath for 30-60 seconds. After air bath, open the next door and enter into the Inoculation Room. The double door system will reduce the entry of pathogen from outside. Inside this room there are a number of sub-components such as Laminar Airflow Cabinet, vacuum pressure machine, etc.

a) Laminar Airflow Cabinet: It is an accessory in Inoculation Room for aseptic manipulation. Explants are sterilized and transferred/inoculated on culture medium under laminar airflow cabinet in this room. The cabinet with horizontal airflow (Fig. 2.3a) from back to front is generally used in tissue culture laboratory. Vertical airflow (Fig. 2.3b) cabinet is used when work with pathogen or Agrobacterium-mediated transformation is done. Laminar airflow cabinet is available in various sizes and shapes to suit different set up of tissue culture laboratory. A motor blows air at a constant velocity of 27 ± 3 m/minute. The constant outward airflow does not permit suspended particles to enter into the working platform. The airflow does not hamper the use of a spirit lamp or gas (LPG) burner. The airflow cabinet has two air filters, a coarse filter which filters large dust particles and subsequently passes through a high efficiency particulate air (HEPA) filter. The appropriate pore size of the HEPA filter is 0.3 μm having 99.97-99.99% filtering efficiency to filter bacterial and fungal spores. The airflow cabinet has two types of lights- UV light to sterilize the microenvironment of the cabinet and fluorescent light which glow during transfer of explant on medium. UV-lights or germicidal lights are turned on at least 30 minutes prior to use the cabinet. Thus an aseptic microenvironment is maintained inside the laminar airflow cabinet. Precautions must be taken to put off the UV-light during inoculation. There are two types of laminar airflow cabinets: (1) Horizontal Laminar Flow Cabinet and (2) Vertical Laminar Flow Cabinet. Horizontal type of laminar is usually being used for tissue culture.

Horizontal Laminar Airflow Cabinet: This equipment discharge HEPA-filtered air from the back of the cabinet across the work surface and towards the user (Fig. 2.3a). This device only provides product protection. They can be used for certain clean activities, such as the dust-free assembly of sterile equipment or electronic devices. Clean benches should never be used when handling cell culture materials or drug formulations, or when manipulating potentially infectious materials. The worker will be exposed to the materials being manipulated on the clean bench potentially resulting in hypersensitivity, toxicity or infection depending on the materials being handled.

Vertical Laminar Airflow Cabinet: Vertical laminar flow cabinet may be useful, for example, in hospital pharmacies when a clean area is needed for preparation of intravenous solutions (Fig. 2.3b). While these units generally have a sash, the air is usually discharged into the room under the sash, resulting in the same potential problems presented by the horizontal laminar flow clean benches. These benches should never be used for the manipulation of potentially infectious or toxic materials.

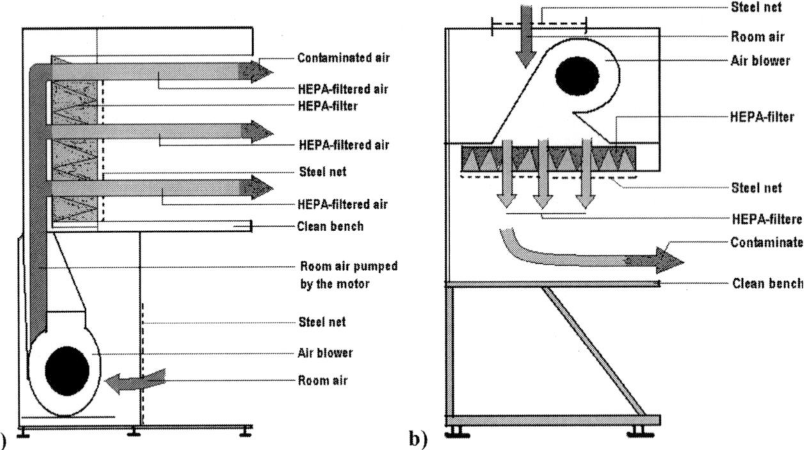


**Fig. 2.3.** Laminar Airflow Cabinet. **a)** Horizontal Laminar Airflow Cabinet, **b)** Vertical Laminar Airflow Cabinet

The platform of the laminar airflow cabinet is sterilized using any surface sterilizing agent, commonly used surface sterilizing agent are 70% rectified spirit, 70% ethanol, SAVLON or 10% calcium chloride etc.

b) Sterilization of Inoculation Room: The air outside the Laminar Airflow Cabinet should be sterilized using germicidal lights. Number of UV-lights depends on the size of the inoculation room. Minimum of two UV-lights on the roof of inoculation chamber are needed at both sides of the Laminar Airflow Cabinet. Here also UV-lights are turned on at least 30 minutes prior to use the cabinet. The floor and side-walls of this room is sterilized using surface sterilizing agents. Once in a week the room can be fumigated. Commonly used fumigating agents are potassium permanganate and acetone. Method: Take about 50 g of potassium permanganate in a glass plate. Keep it at a convenient place so that immediately after pouring acetone on potassium permanganate, you can leave the room. It releases a poisonous gas, hence after uses the room should be kept idle at least for one day.

*2.2.1.2. Dry sterilization*

Glass-wares, small metallic instruments etc. can be sterilized using hot dry air for 2-4 hours at 160-180 °C in a hot air oven. All the items are properly sealed before sterilization with aluminium foil or brown paper or sealed metal container. But, sealing with paper is not advisable as it may get destroyed at higher temperature. An exposure of 160 °C dry hot air is considered as equivalent to steam sterilization at 121 °C and 15 psi for 15 minutes. The un-uniform air circulation and slow penetration of hot air is a major disadvantage in hot air oven sterilization.

Nowadays autoclaves are available with both dry and steam sterilization facilities. When empty glass-wares, small instruments, pipettes, tips are to be autoclaved, use dry cycle of the autoclave.

*2.2.1.3. Sterilization of Laboratory-Ware, Small Instruments and Culture Media*

It is a common practice to use autoclave (steam sterilization) to sterilize glass-wares, media, small instruments and equipments (membrane filter apparatus, etc.) and other accessories such as tissue paper, alluminium foils, filter papers, cotton plugs, distilled water, pipettes, scalpel, spatula, needle etc. All plasticware or plastic materials cannot be sterilized using autoclave under high temperature and pressure. Some types of plastic-wares can also be sterilized in autoclave, such as propylene, polymethyl pentene, pollyalomer, Teflon etc.

Autoclave is a laboratory equipment where water is heated to make vapour. An autoclave has a temperature range of 115-135 °C. Pressure inside the autoclave can also be controlled. There is pressure gauge on the autoclave which indicates the pressure development inside the autoclave.

Empty glass-wares, plastic-wares and other accessories are commonly autoclaved for 15 minutes at 121 °C temperature and 15 psi pressure to completely eliminate the microbes on the surface of these materials. Media in glassware containing different volume are autoclaved for different duration (Table 2.1). Domestic pressure cooker of 22 liter size also can be used for steam or dry autoclaving when small amount of materials (media or other accessories).

**Table 2.1.** Minimum time requirement to make pathogen free nutrient media

| Volume of medium (ml) | Sterilization time (minute) |
|---|---|
| Up to 200 | 15 |
| 200-1000 | 30 |
| 1000-2000 | 40 |

## Changes take place during autoclave

- The pH of autoclaved media is lowered by 0.3-0.5 unit
- Autoclaving at higher temperature can caramelize sugars, which may be toxic to explant
- Autoclaving for longer duration than the prescribed may precipitate the salts and depolymerize agar
- Volatile substances can be destroyed due to autoclaving
- At pressure than the prescribed one, may lead to decomposition of carbohydrates and other components of media
- Vitamins, amino acids, plant extracts, hormones are heat-liable and may decompose during autoclave

**Precautions to be taken during steam sterilization**

- Pour appropriate amount of water in the autoclave/pressure cooker before loading the materials to be autoclaved
- Maintain appropriate pressure and temperature inside the autoclave
- The bottles should not be tightly corked or capped
- Switch off the autoclave immediately after prescribed time duration

*2.2.1.4. Sterilization of Explant*

The part of plant used for tissue culture is termed as ***explant***. Microbes are omnipresent in the ***atmosphere***, even with non-sterilized explants. To obtain sterile plant materials, the explant(s) should be washed with surface-sterilizing chemicals. The commonly used surface-sterilizing chemicals are being given in Table 2.2. and few of them are being briefly narrated below.

### a) Mercuric Chloride ($HgCl_2$)

$HgCl_2$ is used as a topical antiseptic and disinfectant for inanimated and animated (particularly tissue cultural explants) objects. It is used at concentration of 0.1-1.0% for surface sterilization of explant for tissue culture. It is highly toxic. It should be handled with proper care. Keep in a tightly closed container, stored in a cool, dry, ventilated area. Protect from physical damage and direct sunlight. Isolate from incompatible substances. Follow strict hygiene practices. Containers of this material may be hazardous when empty since they retain product residues (dust, solids); observe all warnings and precautions listed for the product.

**Table 2.2.** Explant sterilizing agents, its required concentration and treatment time

| Sterilizing agent | Concentration (%) | Treatment time (min) |
|---|---|---|
| 1. Calcium hypochlorite | 9-10 | 5-30 |
| 2. Mercuric chloride | 0.1-1.0 | 2-10 |
| 3. Sodium hypochlorite | 0.5-5 | 5-30 |
| 4. Silver nitrate | 1.0 | 5-30 |
| 5. Hydrogen peroxide | 3-12 | 5-15 |
| 6. Benzalkonium chloride | 0.01-1.0 | 5-20 |
| 7. Bromine water | 1-2 | 2-10 |
| 8. Elemental iodine | 0.005 | 15-20 |
| 9. Antibiotics | | |
| a) Penicillin | 400 μg | - |
| b) Rifampicin | 25 μg | - |
| c) Streptomycin | 100 μg | - |
| d) Cefotaxime (sodium) | 75-100 μg | - |
| e) Cephalosporin | 150 μg | - |
| f) Benomyl | 50 mg | - |
| g) Nystatin | 25 mg | - |

Whatever cannot be saved for recovery or recycling should be handled as hazardous waste. Processing, use or contamination of this product may change the waste management options. State and local disposal regulations may differ from federal disposal regulations. Dispose of container and unused contents in accordance with local requirements.

### b) Sodium Hypochlorite (NaOCl)

It is one of the most frequently used surface sterilizing agents for plant tissue culture at a concentrations of 0.025-0.25%. Diluted household bleach can also be used for this purpose, which normally contains 5.25% NaOCl. It is equally effective and considerably cheap.

**c) Calcium Hypochlorite (CaOCl)**

CaOCl comparatively cause less damage to plant tissues, but it tends to precipitate in the solution. It can be used in place of NaOCl. To avoid the accumulation of CaOCl precipitations on the plant tissues surface, the solution is to be filtered.

**d) Hydrogen Peroxide ($H_2O_2$)**

$H_2O_2$ is a very pale blue liquid, slightly more viscous than water that appears colourless in dilute solution. It is a weak acid, has strong oxidizing properties, and is a powerful bleaching agent. It is used as a disinfectant, antiseptic and oxidizer. It is easy to use for surface sterilization of plant explant for tissue culture as compared to NaOCl and CaOCl, but it is a bit costly. It is used at concentrations of 3-10%.

**Method**

Collected plant materials are chopped to give proper shape and size to prepare explants. In case of seed as explants, seeds can be dehusked/decoated to prepare explant or can be used as such. Finally prepared explants are washed in liquid detergent (teepol Tween 20, Triton X-100 etc.). Explants are then sterilized under Laminar Airflow Cabinet with sterilizing agent depending up on nature of explant.

*2.2.1.5. Filter Sterilization*

Some chemicals are heat sensitive and its activity may be destroyed during autoclave. Some heat-liable chemicals are amino acids, vitamins, antibiotics etc. These chemicals are sterilized through filter membrane. The pore size of these membranes ranges from 0.45 to 0.22 μm. As the pore size is very small, the chemical (in liquid form) will pass through the membrane in a very slow rate or some time it will not flow down. Here vacuum filter setup is being used. Thus, vacuum machine is essential in a tissue culture laboratory. Most of the filter membranes are of cellulose acetate and/or cellulose nitrate. Some time membrane and assembly units are available separately or preassembled sterilized filter unit of different capacity ranging from 1-200 ml. Preassembled filters are disposable, cannot be reused. In case of unassembled unit, the membrane is properly placed in the assembly unit and it is autoclaved after proper sealing. The unit is open inside the Laminar Airflow Cabinet for filtering the chemical. Filtered chemical is collected in sterilized (autoclaved) container. This filtered chemical(s) can be added in the melted medium under Laminar Airflow Cabinet and mixed properly.

*2.2.1.6. Flame Sterilization*

Burner is an important component inside the Laminar Airflow Cabinet. It keeps pathogen free with its radius of 15 cm. Forceps, spatula, scalpel, and other metal apparatus are sterilized by burning on the flame. These tools are dipped in alcohol and burnt to red hot on flame. In place of flame burning, nowadays, these apparatus can be sterilized in hot glass beads chamber which run by electricity.

## 2.3. RULES OF CONDUCT IN TISSUE CULTURE LABORATORY

In any laboratory there will be lot of chemicals of all ranges, instruments and equipments which should be handled with care and in a proper way. If it is not so, we may face some serious accidents. Tissue culture laboratory is not the exception; every worker in the laboratory ought to follow all the rules applied in each component of Tissue Culture Laboratory.

### 2.3.1. Precautions

The precautions to be followed are being listed below component-wise.

*2.3.1.1. Cleaning Room*

- Always use gloves while washing laboratory-wares.
- Broken glass-wares should be collected immediately and to be disposed at a safe place.
- Contaminated laboratory-wares should be autoclaved before bringing to washroom.
- Avoid rinsing of plastic-wares with organic solvents, such as acetone, chloroform or methyl chloride.
- Remove labels and marking ink before soaking in detergent solution

*2.3.1.2. Autoclave Room*

- Before loading materials to be autoclaved, check whether there is sufficient water in the autoclave chamber.
- Adjust the pressure of the autoclave as per requirement (15 psi).
- Switch off the autoclave immediately after the prescribed time is over.
- It is advisable not to stay in the autoclave room when it is on to avoid unforeseen accident. Few cases of bursting of autoclave have been reported in different laboratories and institutes.

#### *2.3.1.3.. Media Preparation Room*

- The measuring cylinder used to prepare stock/medium should be rinsed in tap water after use.
- Immediately after dispensing medium from beaker/conical flask to culture tubes/ Petri-dishes, the beaker should be washed in tap water to avoid sticking of agar to the walls of laboratory-ware.
- When preparing a solution, add the individual chemical in proper order.
- Do not leave any chemical on the working table, immediately after its use keep it at its own place.

#### *2.3.1.4. Inoculation Room*

- Keep a thick cotton/woolen towel on the Laminar Airflow Cabinet. During the process of explant transfer or some other work inside the Laminar Airflow Cabinet, spirit or alcohol may spill out and catch fire. Put the towel on the flame to put off the fire.
- Wear cotton apron during inoculation and transfer of explant on culture medium.
- Before starting your work, confirm that UV light of inoculation room and inside the Laminar Airflow Cabinet is switched off.
- Keep your hand at safe distance from the flame of burner, as you have sterilized your hand with spirit, it may catch fire.

#### *2.3.1.5. Culture Room*

- Never fumigate in the Culture Room when there is plant culture.
- Do not use UV-light inside Culture Room.
- Regularly surface-sterilize the floor, walls and racks of Culture Room.
- Temperature of culture room should be maintained at 25 ±°C.
- The light intensity should be maintained as per requirement.

### 2.3.2. Material Safety Data Sheet

A material safety data sheet (MSDS) is a form containing data regarding the properties of a particular substance. An important component of product stewardship and workplace safety, it is intended to provide workers and emergency personnel with procedures for handling or working with that substance in a safe manner, and includes information such as physical data (melting point, boiling point, flash point, etc.), toxicity, health effects, first aid, reactivity, storage, disposal, protective equipment, and spill handling procedures. The exact format of an MSDS can vary from source to source within a country depending on how specific is the national requirement.

MSDSs are widely used system for cataloging information on chemicals, chemical compounds, and chemical mixtures. MSDS information may include instructions for the safe use and potential hazards associated with a particular material or product.

There is also a duty to properly label substances on the basis of physico-chemical, health and/or environmental risk. An MSDS for a substance is not primarily intended for use by the general consumer, focusing instead on the hazards of working with the material in an occupational setting. For example, an MSDS for a cleaning solution is not highly pertinent to someone who uses a can of the cleaner once a year, but is extremely important to someone who does this in a confined space for 40 hours a week.

In some jurisdictions, the MSDS is required to state the chemical's risks, safety and impact on the environment.

It is important to use an MSDS that is both country-specific and supplier-specific as the same product (e.g. paints sold under identical brand names by the same company) can have very different formulations in different countries.

## 2.4. CONCENTRATION OF SOLUTIONS

For preparation of medium, stock solutions and different growth regulators etc. defined in concentration. ***Concentration*** of a solution is the relative proportion of the solute in relation to solvent. The concentration of chemical composition of various components of a solution or synthetic medium can be expressed in various units, such as weight, equivalent weight, percentage, molarity, normality etc., which are briefly discussed below:

### 2.4.1. Atomic Weight

Atomic weight of an element is the relative weight of the atom. The IUPAC definition (IUPAC, 1980) of atomic weight is "An atomic weight (relative atomic mass) of an element from a specified source is the ratio of the average mass per atom of the element to $1/12^{th}$ of the mass of an atom of $^{12}C$".

### 2.4.2. Molecular Weight

The sum of the atomic weight of all the atoms in a molecule is termed as molecular weight. The unit of weight is the dalton, one-twelfth the weight of an atom of $^{12}C$. Thus the molecular weight (MW) of water is 18 daltons (we shall ignore the tiny error introduced by the presence of traces of other isotopes - $^{17}O$, $^{18}O$, and $^{2}H$ among the predominant $^{1}H$ and $^{16}O$ atoms). For example, the molecular weight of $CaCl_2.2H_2O$

$$= 40.08 + (2 \times 35.453) + (4 \times 1.008) + (2 \times 16)$$

$$= 147.018$$

### 2.4.3. Equivalent Weight

The equivalent weight is the weight of a substance that will combine with or replace one mole of hydrogen or one-half mole of oxygen. The equivalent weight is equal to the atomic weight divided by the valence. With reference to chemistry, equivalent weight is the weight of an element or compound that will combine with or displace 8 grams of oxygen or 1.00797 grams of hydrogen. Also called gram equivalent. The molecular weight of an acid is the weight of that contains one atomic weight of acid hydrogen, that is, that hydrogen that reacts during neutralization of acid with base. For example, the equivalent weight of $H_2SO_4$ is 49, since $H_2SO_4$ contains two replaceable hydrogen.

$$\text{Equivalent weight} = \frac{\text{Molecular weight}}{\text{Valency of cations or anions (hydrogen equivalent)}}$$

### 2.4.4. Weight

Weight is a measure of the heaviness of an object. The force with which a body is attracted to Earth or another celestial body, equal to the product of the object's mass and the acceleration of gravity.

- It is usually represented as g/L or mg/L for preparation of solution
- $10^{-6}$ = 1.0 mg/L

  = 1 ppm (part per million)
- $10^{-7}$ = 0.1 mg/L
- $10^{-9}$ = 0.001 mg/L

= 1 µg/L

### 2.4.5. Percentage

The dictionary meaning of percentage (%) is a rate or number or amount in each hundred. It may be expressed in w/w or w/v or v/v.

Where, w = weight, v = volume

### 2.4.6. Parts per million

Number of share of million parts of a solution or mixture is termed as parts per million (ppm). That is, one ppm solution contains one part of the substance in million parts of the solution or mixture. It may also be expressed in w/w or w/v or v/v.

1 mg in 1 kg = 1 ppm

1 mg in 1 L = 1 ppm

1 mL in 1 L = 1 ppm

### 2.4.7. Molar (M) Concentration

A mole is the quantity of a substance whose weight in grams is equal to the molecular weight of the substance. One molar concentration of solution contains the same number of gram of substance as is given by its molecular weight in one liter of solution. Thus 1 mole of glucose weighs 180 g. Furthermore, if you dissolve 1 mole of a substance in enough water to make 1 liter (L) of solution, you have made a 1-molar (1 ***M***) solution.

1 M = the molecular weight in g/L

1 mM = the molecular weight in mg/L

= $10^{-3}$ M

1 µM = the molecular weight in µg/L

= $10^{-6}$ M

= $10^{-3}$mM

### 2.4.8. Normal (N) Solution

One normal solution of a substance contains one equivalent or one gram equivalent weight of the substance in one liter of solution.

### 2.4.9. Conversion from milli molar (mM) to mg/L

For example, the molecular weight of sucrose is 342 g.

1 M sucrose solution consists of 342.0 g/L

1 mM sucrose solution consists of 0.342 g/L or 342 mg/L

1 µM sucrose solution consists of 0.000342 g/L or 0.342 mg/L

### 2.4.10. Conversion from mg/L to mM

For example, the molecular weight of 2, 4-D is 221 g

If 6 mg/L is to be converted into mM, then

$$\text{The number of mM 2, 4-D} = \frac{\text{No. of mg 2, 4-D}}{\text{Molecular weight of 2, 4-D}}$$

$$= \frac{6}{221} = 0.027 \text{ mM}$$

*2.4.1.11. Preparation of Solution*

Prepare 200 ml sucrose solution of 12 mg/L concentration from already prepared sucrose solution, whose concentration is 35 g/L.

We know, $C_1V_1 = C_2V_2$

Where, $C_1$ = Concentration of solution 1 (12 g/L)

$V_1$ = Volume of solution 1 (200 ml)

$C_2$ = Concentration of solution 2 (35 g/L)

$V_2$ = Volume of solution 2 (??)

$$\text{Then, } V_2 = \frac{C_1V_1}{C_2}$$

$$= \frac{12 \times 200}{35} = 68.57 \text{ ml}$$

## 2.5. COMPONENTS OF NUTRIENT MEDIA

Plants in nature are ***autotrophic***, can synthesize their own food material using $H_2O$ and $CO_2$ in presence of light. In contrast, plants growing in vitro are ***heterotrophic***. They cannot synthesize their own food material. Plant tissue culture media therefore require all essential minerals plus a carbohydrate source usually added in the form of sucrose and also other growth ***hormones*** (regulators and vitamins). Growth and ***morphogenesis*** of plant tissues in vitro are largely governed by the composition of the culture media. Although the basic requirements to culture plant tissues are similar to those of plants grow in natural conditions. In practice, nutritional components promoting optimal growth of a tissue under laboratory conditions may vary in respect to the particular species. Media compositions are thus formulated considering specific requirements of a particular culture system. During the course of evaluation of plant tissue culture media, it was noted that the composition, form of nutrient salts as well as the concentration of the nutrients is very important.

### 2.5.1. Momentary History of Plant Tissue Culture Media

German botanist Gottlieb Haberlandt (1902) developed the concept of in vitro plant cell culture. Thus he is regarded as the father of plant tissue culture. Considerable progress has been made since 1950 on the development of media for growing plant cells, tissues and organs aseptically in culture. A significant contribution to formulation of a defined growth medium suitable for a wide range of applications was made by Murashige and Skoog (1962) - MS. The MS medium was developed for culture of tobacco explants and was formulated based on an analysis of the mineral compounds present in the tobacco itself. MS medium contains relatively high salt levels, particularly potassium and nitrogen. In their work to adapt tobacco callus cultures for use as a hormone bioassay system, they evaluated many medium constituents to achieve optimal growth of calluses. In so doing, they improved upon existing types of plant tissue culture media to such an extent that their medium (the MS medium) has since proved to be one of the most widely used in plant tissue culture work that gives the composition of different media and it becomes the most cited reference in the science of plant tissue culture.

Later, this MS medium was modified by Linsmaier and Skoog (1965) - LS. Meanwhile, White (1963) developed a medium (White's medium) with a low salt formulation for culture of tomato roots. Gamborg and coworkers (Gamborget al., 1968) composed B5 medium for soybean callus culture and contains a much greater proportion of nitrate compared to ammonium ions. Nitsch and Nitsch (1969) developed medium for anther culture and it contains lower salt concentrations than that of MS medium but not as low as that of White's medium. Another medium, designated as $N_6$ (Chu, 1978) has been developed especially for cereal anther culture.

Virtually all tissue culture media are synthetic or chemically defined compounds; only a few of them use complex organics, e.g., potato extract, as their normal constituents. A variety of recipes have been developed since none of them is suitable for either all plant species or for every purpose. Most of these recipes have been elaborated from those of White (itself evolved from a medium for algae) and Gautheret (based on Knop's salt solution). The composition of White, MS and B5 (the last two are the most commonly used) N6, Nitsch's, White and Gautheret media are being presented in Table 2.5.

### 2.5.2. Components of Synthetic Media

The standard protocols for plant tissue culture can be altered as per the requirement. Working with new species, initially different media can be tested and growth conditions must be determined empirically. Every plant tissue culture medium has a common essential nutrient component to support in vitro plant culture as listed below.

a) Inorganic nutrients
- Macronutrients
- Micronutrients

b) Carbon as energy source

c) Organic supplements
- Vitamins
- Amino acids
- Other organic supplements
- Activated charcoal
- Antibiotics

d) Growth regulators

e) Solidifying agents

#### *2.5.2.1. Inorganic Nutrients*

Plant growth and development can proceed only when the plants are supplied with the chemical elements referred as ***essential elements***. Naturally plants are ***autotrophic*** (self-feeding) and they absorb nutrients by the roots from the soil. Since the sources of these inorganic requirements are minerals, these elements are known as mineral nutrients. Chemical analysis of plant ash showed that plants contain about 40 different elements. Some of them are indispensable or necessary for the normal growth and development of plants and they are known as ***essential elements***. Rests of them are called ***non-essential elements***. It is known that 15 elements are essential for majority of the plants, they are- carbon, hydrogen, nitrogen, oxygen, phosphorus, potassium, calcium, sulphur, magnesium, iron, zinc, boron, copper, manganese, and molybdenum.

In case of in vitro culture, the nutrients are supplied in the form of salts. A variety of mineral salts supply the needed elements for normal growth and development of plant. Unorganized tissue cultured explant or regenerated organ or incomplete plantlets are ***heterotrophic***. These plants need supply of organic and inorganic substances for the proper growth and development. The synthetic medium should contain the correct amount and proportion of inorganic nutrients to satisfy the nutritional as well as physiological needs of the many plant cells in culture. Based on the requirement of the elements, the inorganic nutrients can be classified into two groups, namely macronutrients and micronutrients.

**a) Macronutrients**

The elements required comparatively large in quantity, that is, in concentrations more than 0.5 mmol $L^{-1}$ (or 30 ppm) are known as ***macronutrients***. These include six elements- nitrogen (N), phosphorus (P), potassium (K), calcium (Ca), magnesium (Mg) and sulphur (S). These macronutrients provide both cation and anion for the growth and development of plant cell. These elements have both structural and functional roles in protein synthesis (particularly N and S), nucleotide synthesis (P, N and S), cell wall synthesis (Ca), enzyme cofactors (Mg) and membrane integrity (Mg). For better result, inorganic nitrogen is used in both the forms- nitrate and ammonium. Nitrate is used in the range of 25-40 mM and ammonium in the range of 2-20 mM in the synthetic medium. Most plant prefer nitrate to ammonium, although the opposite is also true in some cases.

Potassium is important for ion balance in the cells. The medium should contain at least 25 mM $L^{-1}$ of potassium. The concentration of K in cytoplasm is high (100-200 mM) and in the chloroplast (20-200 mM). Potassium salt maintains osmotic pressure of cytoplasm and involve in osmotic regulation of cell. It acts as an activator of many enzymes involved in carbohydrate metabolism and protein synthesis. It is actively involved in the opening and closing of stomata. Phosphorus is another important essential element and it is involved in energy metabolism of cell. It is an essential element participating in the skeleton of plasma membrane, nucleic acids and organic molecules such as ATP and other phosphorylated compounds. P is available in the tissue culture medium as sodium hydrogen phosphate or potassium hydrogen phosphate.

Other macronutrients, Ca, S and Mg present in the synthetic medium in the range of 1-3 mmol $L^{-1}$. Calcium is the important constituent of middle lamella in the cell wall and essential in the functioning of cell membranes. It helps to stabilize the structure of chromosome. Ca is also concerned with the growing root apices. In the tissue culture medium it is supplied in the form of calcium chloride and calcium nitrate. Magnesium is very immobile and essential for many enzymatic reactions. It is a constituent of chlorophyll and it is indispensable for photosynthesis. Mg is essential for energy metabolism of the plant, particularly in ATP synthesis. It is available in the tissue culture medium as magnesium sulphate. Sulphur is another important macronutrient of plant. It is an important constituent of some amino acids (cystine, cysteine and methionine), vitamins (biotin, thiamine), co-enzyme A and volatile oils. In synthetic medium, it is supplied in the form of magnesium sulphate and potassium sulphate.

**b) Micronutrients**

The essential elements, which required in the concentrations less than 0.5 mmol $L^{-1}$ is known as ***micronutrients***. These include iron (Fe), zinc (Zn), boron (B), copper (Cu), manganese (Mn) and molybdenum (Mo). Requirement of some of the micronutrients are not yet clear, they are still essential. Apart from these elements, some more elements also have been identified and realized their necessity for the normal growth of some plants, for example, chlorine (Cl) for coconut, silicon (Si) for rice, cobalt (Co) for groundnut, iodine (I), and sodium (Na). The actual role of these elements in plant growth and development is yet to be established.

Among the micronutrients, Iron (Fe) is more critical as chelated form of iron is commonly used in plant tissue culture medium. Due to its solubility problem, it is supplied in the plant tissue culture medium with a chelating agent – sodium or potassium form of ferric ethylene diamine tetra-acetic acid ($Na_2$Fe EDTA). ***EDTA*** helps iron to be available in the culture. In this form iron remains available up to pH 8.0. Fe is important constituent of iron-porphyrin proteins such as cytochromes and also for enzymes like peroxidases, catalases etc. It is essential for synthesis of chlorophyll. It acts as a catalyst and electron carrier during respiration.

Zinc is supplied in the medium in the form of zinc sulphate. It acts as metal component of a number of enzymes. It plays an active role in synthesis of protein, tryptophan and chlorophyll. It is also essential for the biosynthesis of plant growth hormones - IAA. Manganese is added to the synthetic medium in the form of manganese sulphate. Mn is an important element for ***Hill reaction*** of photosynthesis. It acts as activator of respiratory enzymes like oxidase, peroxidase, dehydrogenase, kinase, decarboxylase etc. Boron is an essential nutrient, involves in sugar translation, protein synthesis and seed cell wall formation. It is added in the medium as boric acid. Copper plays an important role in photsynthesis, particularly in the intermediate of the electron transport chain between photosynthesis. It is also essential to the activity of various enzymes, for examples, phenolase, ascorbic acid oxidase etc. It is supplied in the medium in the form of copper sulphate. Molybdenum is present in the artificial medium as sodium molybdate. Mo participates in the process of conversion of nitrate into ammonium.

*2.5.2.2. Carbon and Energy Source*

In vitro cultures lack in autotrophic ability, which requires supply of carbon as external energy source. Sometimes, the green callus does not also have the autotrophic ability. Sucrose has been considered the most effective carbohydrate source which can not be substituted by other disaccharides. The addition of carbon sources in the tissue culture medium enhances cell proliferations and greening or morphogenesis of cultured cells on artificial medium. On autoclave, sucrose converts into glucose and fructose. The cultured cells first use glucose and then fructose. Better callus growth followed by regeneration was observed when autoclaved sucrose is used in the medium than the filtered sterilized sucrose. This suggests that cells benefit from the readily available supply of glucose and sucrose brought about by hydrolysis of autoclaved sucrose. It is generally used at a concentration of 2-6%.

Other carbon sources used in plant tissue culture media are maltose, lactose, galactose, rafinose, cellobiose, melibiose, trehalose and starch. These were found to be much inferior to sucrose. However, use of maltose in anther culture was found to produce better result than sucrose (Roy and Mandal, 2005).

*2.5.2.3. Organic Nutrients*

Organic nutrients include vitamins, amino acids, and other organic supplements, such as, potato extract, coconut milk, orange juice, tomato juice, beef extract, yeast extract, amino acids, casein hydrolysate etc.

**a) Vitamins**

In nature, plants endogenously synthesis all vitamins required for growth and development. Vitamins are used as catalysts in various metabolic processes. Under in vitro culture conditions, plant tissues produces some vitamins but at a sub-optimal quantities. Hence, the exogenous supply of vitamins is essential for normal growth and development of tissues or callus or plantlets. The most frequently used vitamins are-thiamine ($B_1$), nicotinic acid ($B_3$), pyridoxine ($B_6$). Other vitamins that are generally used in plant tissue culture medium are- p-Aminobenzoic acid, Folic acid, Choline chloride, Biotin, Riboflavin, Retinol (vitamin A), Tocophenol (vitamin E), Nicotinamide, Ascorbic acid, Calcium pentothenate ($B_5$) Vitamin $B_{12}$, and Vitamin $D_3$. These vitamins are added to the medium at a range of 0.1-10.0 mg $L^{-1}$. Thiamine is used by all the plant cells under in vitro for normal growth and development.

**b) Amino acids**

In vitro cultured cells are capable of synthesizing amino acids required for various metabolic processes. These nitrogen containing amino acids augment nitrogen supply, enhance cell growth and facilitate regeneration. The most frequently used amino acid in different plant tissue culture media is glycine. Other commonly used amino acids are casein hydrolysate, L-glutamine, L-asparagine, L-arginine, and L-cysteine. Sometimes amino acid is inhibitory to plant cell growth, thus it should be used very carefully.

**c) Antibiotics**

Antibiotics are substances produced by certain microorganisms that suppress the growth of other microorganisms and eventually destroy them. Antibiotics are used when explants have a ***systemic*** or ***symbiotic*** association of micro-organisms (especially bacteria or fungus). Their applications include:

- Suppresses bacterial infections in plant cell and tissue culture.
- Suppresses mould and yeast infections in cell cultures.
- Eliminates **Agrobacterium** species after the transformation of plant tissue.

Generally, addition of antibiotics in tissue culture medium is avoided as it reduces the growth of cells and tissue. They are usually dissolved in distilled sterile water or a convenient solvent, filter sterilized, then added to the autoclaved medium cooled to 45-50 °C. Commonly used antibiotics are kanamycin, streptomycin, hygromycin etc. Some other antibiotics which can be used for plant tissue culture are being given in Table 2.3.

**d) Sugar alcohol**

The commonly used sugar alcohol or hexitol is myo-inositol. It helps in sugar transport, carbohydrate metabolism, membrane structure and cell wall formation. Manitol and sorbitol are two sugar alcohols which are used to regulate osmotic potential in ***protoplast*** culture. Manitol and sorbitol are nutritionally inert.

### e) Additive for explants which release phenolic/phenol-like compound

Tissues injured during explant excision from the stock plants or during preparation and sizing the explant often cause release of various compounds that are air oxidized (by peroxidases or polyphenoloxidases) and turn brown or black resulting darkening both tissues and medium. Sometimes, these are toxic to culture, which may inhibit culture growth or even may kill the culture.

**Table 2.3.** Antibiotics usually used in plant tissue culture and their mode of action

| Antibiotic | Mode of action |
|---|---|
| **Amino glycosides** | Manamycin, Neomycin, Streptomycin |
| Chloramphenicol | |
| **β-Lactams** | Chloramphenicol |
| Ampicillin, Carbenicillin, Penicillin | |
| **Chloramphenicol** | Inhibits protein synthesis by acting on 50S ribosome |
| **Glycopeptides** | Interfere with bacterial cell wall synthesis |
| Chloramphenicol | |
| **Macrolides and lincosamides** | Inhibit protein synthesis by acting on 50S |
| Erythromycin, Lincomycin | ribosome |
| Polymixinns | |
| **Polymixinns** | |
| Polymixinns B, E | Attach to the cell membrane and modify ion flux, resulting in cell lysis |
| **Quinolones** | |
| Nalidixi acid, Norfloxacin, Ofloxacin | Interfere with DNA replication by inhibition of DNA gyrase |
| **Rifampicin** | Interferes with mRNA formation by binding to RNA polymerase |
| **Tetracyclines** | Inhibits protein synthesis by acting on |
| 30S | ribosome |
| **Trimethoprim and Sulphonamides** | Inhibit synthesis of tetrahydrofolate |

### f) Activated charcoal

Activated charcoal (AC), when incorporated into artificial seeds, improved the conversion and vigour of the encapsulated propagules of tropical forest trees by not only stimulating the diffusion of gases and nutrients, but also by helping in breaking down alginate which in turn facilitates enhanced respiration of propagules, thus preventing the loss of vigour that extends the storage period significantly (Saiprasad,

2001). AC also absorbs unintended and unwanted exudates such as 5-hydroxymethylfurfural (a toxic breakdown product of sucrose formed during autoclaving) and other harmful phenolic products (Wang et al., 2007). AC also retains nutrients within the hydrogel capsule and releases them slowly, thus providing a long-term supply of nutrients to the growing tissue. AC at 1.25% (w/v) improved the conversion frequency of encapsulated somatic embryos in *Oryza sativa* (rice) (Arun Kumar et al., 2005).

Activated charcoal has a very fine network of pores with large inner surface area on which many substances can be adsorbed. Activated charcoal is often used in tissue culture to improve cell growth and development. It plays a critical role in micropropagation, orchid seed germination, somatic embryogenesis, anther culture, synthetic seed production, protoplast culture, rooting, stem elongation, bulb formation etc. The promontory effects of activated charcoal on morphogenesis may be mainly due to its irreversible ***adsorption*** of inhibitory compounds in the culture medium and substantially decreasing the toxic metabolites, phenolic exudation and brown exudates accumulation. In addition to this, activated charcoal is involved in a number of stimulatory and inhibitory activities including the release of substances that are naturally present in activated charcoal which promote growth, alternation and darkening of culture media, and adsorption of vitamins, metal ions and plant growth regulators, including absicic acid and gaseous ethylene.

If the explants release ***phenolic compounds*** in the medium, activated charcoal is frequently used at a concentration of 0.2-5.0% (w/v). It helps in reduction of toxic effects of phenolic compounds produced during in vitro culture and facilitates growth of culture. It helps in removal of inhibitors from the agar used for gelling the medium. Another role as

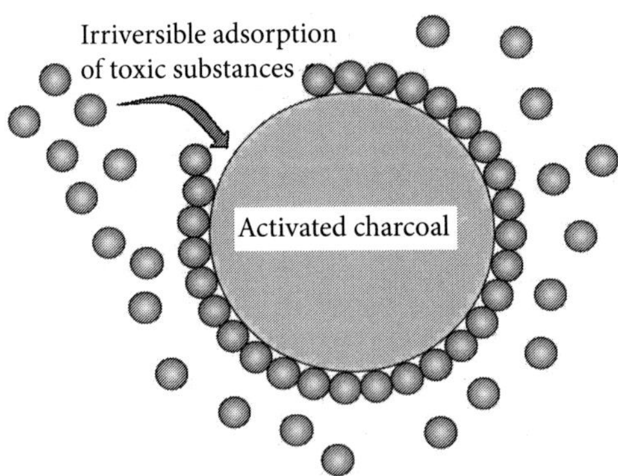


**Fig. 5.4.** Adsorption of toxic substances in the medium by activated charcoal

signed to activated charcoal is the adsorption of 5-hydroxymethylfurfural, a product of sucrose dehydration during autoclaving, assumed to be an inhibitor of growth of anther cultures. It also helps to stabilize the pH of the medium as well as promotes growth and differentiation of some plant species, for example, orchids, carrot, ivy and tomato.

Activated charcoal is generally acid washed and neutralized before its use. Polyvinylpyrroline (250-1000 mg $L^{-1}$), citric acid (100 mg $L^{-1}$) and ascorbic acid (100 mg $L^{-1}$), thiourea, L-cysteine are also used to prevent oxidation of phenolics. These compounds, particularly activated charcoal should be added to the medium only when it is needed, otherwise it may inhibit the growth of cells/ callus by adsorption of phytohormones.

**g) Coconut milk and coconut water**

Coconut water is the liquid endosperm which develops in a free nuclear condition without forming cell wall. Coconut water contains a number of different chemical substances, some of which are active by themselves, some of which interact with non-active compounds present in the endosperm, and some compounds that apparently have no specific growth effect but have nutritive roles (sugars, sugar alcohols, amino acids, amides etc.) Carbohydrates constitute up to 85% of the dry weight of coconut milk, with glucose as the major sugar. The bulk of the carbohydrate fraction is sorbitol, a six carbon alcohol. It also contains several growth promoting substances, namely IAA, gibberellins, and cytokinins.

It was found that the coconut milk support the growth of tissue cultured plant materials in vitro. Thus, many workers used coconut milk as a source of nutrient for culturing embryos and other tissues and organs at a rate of 5-20%. It was reported that coconut water at rate of 15% improves the green plantlet regeneration in rice anther culture (Roy and Mandal, 2004 & 2005). Swamy et al. (2009) successfully used coconut water for artificial seeds conversion.

**h) Beef extract**

It is derived from the fresh beef of healthy animals. Beef extract powder is used for preparing microbiological culture media and some extent of plant tissue culture medium. The function of these beef extract products can be described as complimenting the nutritive properties of peptone by contributing minerals, phosphates and energy sources etc. Beef extract powder provides nitrogen, amino acids, vitamins and carbon. Its concentration is normally in the range of 0.3 to 1.0% in culture media that may vary depending upon the need. However, usually the concentration does not exceed 0.5%.

### i) Other Organic Supplements

Other commonly used organic extracts are protein hydrolysate, (casein-hydrolysate, lactalbumin-hydrolysate, peptone), fish emulsion, yeast extract, potato extract, malt extract, tomato and orange juices, banana pulp etc. Potato extract has been found suitable for anther culture.

### *2.5.2.4. Plant Growth Regulators*

Plant hormones are organic compounds that elicit a physiological response at very low concentrations (generally < 1 mM and often < 1 μM). Plant hormones are also known as ***phytohormones*** or ***plant growth regulators***. Nowadays, it is more correctly termed as plant growth regulator. Growth and development can be influenced by several growth regulators. These growth regulators can be divided into five major groups- auxin, cytokinins, gibberellins, abscisic acid and ethylene (Table 2.4). Ethylene is not used in plant tissue culture. However, some other substances like polyamines, jasmonates, brassinosteriods and salicylic acid are being studied and these may eventually be classified as hormones too. Thesephytohormones are the most critical constituents of synthetic medium.

In vitro cultured plant cells have certain levels of their indigenous plant hormones, which is not sufficient for morphogenesis. The ratio of hormones required for root and shoot induction varies considerably with tissues as well as plant species. Again, the organogenesis depends on the ratio of auxin and cytokine in as well as the indigenous level of the phytohormones. The exogenous addition of plant growth regulator in the nutrient medium increases the level within the tissue. The effects of these growth regulators depend on the rate of uptake, stability of the medium and sensitivity of the particular tissue.

### a) Auxins

Auxins are required by all plants for the normal growth and development. In artificial medium, it is added to induce cell division and callus formation, cell elongation, swelling of tissues, and some time organogenesis (formation of adventitious root). Naturally occurring auxins are 1H-indol-3-acetic acid (IAA) and 1H-indol-3-butyric acid (IBA). IAA should be stored in amber bottle as they are unstable in light. Many synthetic auxins are also used in plant tissue culture media, for example,

1-naphthaleneacetic acid (NAA), 2,4-dichlorophenoxyacetic acid (2,4-D), 2,4,5-trichlorophenoxyacetic acid (2,4,5-T), naphthoxyacetic acid (NOA), p-chlophenoxyacetic acid (pCPA), 4-chlophenoxyacetic acid (4-CPA), 2-methyl-4-chlophenoxyacetic acid (MCPA), 4-amino-3,5,6-trichloropicolinic acid (picloram) and 3,6-dichloro-2-methoxybenzoic acid (dicamba).

Commonly used auxins for callus induction are 2,4-D, NAA, IBA, pCPA, picloram and dicamba. At low concentrations some auxins promote adventitious roots, whereas at high concentration it promotes callus induction and callus growth. The functions of the growth regulators added in the synthetic medium also influenced by the indigenous concentrations of the auxins present in the explants.

**b) Cytokinin**

Cytokinins are the most complex class of growth substances. These are the derivatives of adenine and are mainly $N_6$-substituted amino purines. These growth regulators are associated with cell division, differentiation, embryogenesis and shoot formation (see Table 2.4). Cyntokinins are available in both natural and synthetic forms. Often used cytokinins in plant tissue culture medium are 6-benzylaminopurine (BAP), 6-benzyladenine (BA), N-(2-furfurylamino)-1-H-purine-6-amine (kinetin), $N^6$-($\beta^2$-isopentenyl) adenine (2-iP) and 6-(4-hydroxy-3-methyl-trans-2-butanylamino) purine (zeatin). Zeatin is naturally occurring cytokinin extracted from Zea mays, while BA and kinetin are synthetically derived cytokinin. Kinetin is stored in amber bottle since they are unstable in light.

Some non-purine stable chemicals are also acting as cytokinin having very high activities, namely thidiazuron (TDZ) and N-2-chloro-4-puryl-N-phenyl urea (CPPU) and other derivatives of urea (Pierik, 1989). TDZ has now emerged as highly efficient growth regulator in plant tissue culture. It has diverse effects on plant tissue culture. It influences the callus induction and somatic embryogenesis. TDZ is an interesting growth regulator, which can act both as auxin as well as cytokinin.

The ratio of auxin and cytokinin is the most important for morphogenesis of callus. High concentration of cytokinin and low concentration of auxin in combination in the culture medium promotes shoot formation, whereas the reverse is suitable for callus initiation and proliferation. Both auxin and cytokinin are thermostable. Usually cytokinins are dissolved in 1 N HCl or 1N NaOH whereas zeatin dissolve in ethanol (Table 2.6).

### c) Gibberellins

There are about 90 gibberellins. Among these, $GA_3$ is most commonly used in plant tissue culture. It promotes growth of cultured cells, stem elongation, embryo maturation, but it can inhibit callus growth (Table 2.4). Gibberellins generally inhibit adventitious root and shoot formation. It dissolves in alcohol (Table 2.6). It is not thermostable, so it is advisable to use filter-sterilized gibberellins in plant tissue culture media.

### d) Abscisic acid

Abscisic acid (ABA) inhibits shoot growth, but it enhances germination of embryo and embryo maturation under in vitro culture (Table 2.4). ABA in culture medium either stimulates or inhibits callus growth depending on the species, thus it is used as growth retardant. It is thermostable but light sensitive. Usually filter-sterilized ABA is used in the culture media.

### e) Other plant growth regulators

In general, ethylene is undesirable in plant tissue culture. It is produced by cultured cells, but its role in tissue culture is still unknown. Auxin may promote the ethylene production during culture and it accumulates at the headspace of culture tubes, which may be harmful to cultures. Some other growth regulators or plant growth substances are available, which have their effects on plant growth and developments. These are polyamines, jasmonate, brassinosteriods and salicylic acid. Polyamines are not true growth regulators their presence in the synthetic medium enhance regeneration of roots, shoots and embryo in some plant species. Jasmonate acts both growth enhancer as well as inhibitor. It promotes senescence, abscission, tuber formation, pigment formation, tissue differentiation and adventitious root formation. Brassinosteriods are new class of plant growth substances. They promote shoot elongation, but inhibit root growth and development. Salicylic acid is also considered as a potential plant growth regulator. It has effects on many physiological processes in plants.

**Table 2.4.** Effects of plant growth regulators in plant tissue culture

| Classes | Forms | Effects on plant tissue culture |
|---|---|---|
| Auxin | IAA, IBA, NAA, 2,4-D, 2,4,5-T, Picloram, Dicamba, CPA, MCPA | • Callus formation and growth<br>• Adventitious root formation<br>• Adventitious shoot formation (at low concentration)<br>• Cell division<br>• Differentiation of vascular tissue<br>• Usually inhibition of outgrowth of axillary buds<br>• Usually inhibition of root growth |
| Cytokinin | BA, BAP, kinetin, zeatin, TDZ, 2-ip, CPPU, DPU | • Stimulation of shoot initiation/bud formation<br>• Inhibition of root formation<br>• Cell division<br>• Callus formation and growth (in some cases)<br>• Promotion of rejuvenation of mature shoots<br>• Stimulation of outgrowth of axillary buds<br>• Inhibition of shoot elongation<br>• Inhibition of leaf/shoot senescence<br>• Promotion of some stages of root development |
| Gibberellins | $GA_3$ is most frequently used | • Promotion of internode elongation<br>• Loss of dormancy in somatic embryos, apical buds<br>• Inhibition of adventitious root formation<br>• Regulation of the transition form juvenile to adult phase<br>• Synthesis of inhibitors which facilitate ***acclimatization*** |
| Abscisic acid | - | • Maturation of somatic embryos<br>• Promotion of desiccation tolerance of somatic embryos<br>• Promotion of accumulation of storage protein during embryogenesis<br>• Inhibition of precocious germination of somatic embryos<br>• Inhibition of shoot elongation |
| Polymines | Putrescine<br>Spermidine<br>Spermine | • Promotion of adventitious root formation<br>• Promotion of shoot formation<br>• Promotion of somatic embryogenesis |
| Brassinosteroids | - | • Promotion of shoot elongation<br>• Inhibition of shoot and root elongation (both effect can be noted)<br>• Enhancement of xylem differentiation |
| Salicylic acid | | • Promotes bud formation<br>• Induction of flowering in plants |

When explants are incubated in the dark, they induce more root primordial followed by the addition of PGRs in the gel matrix for higher conversion from encapsulated beads (Piccioni, 1997). Pretreatment of explants with cytokinin(s) and auxin(s) has enhanced the conversion frequency of artificial seeds into plantlets for the following: mulberry (*Maluspumilla*) (Pattnaik et al., 1995); pineapple (*Ananascomosus*) (Soneji et al., 2002); *Dalbergiasissoo* (Chand and Singh, 2004); Carrizo citrange (Germanà et al., 2011); Corymbia spp. and African mahogany (Hung and Trueman, 2012). Moreover, when PGRs are added to the gel matrix, the efficiency of synthetic endosperm around the vegetative propagules improves and provides a simpler method for the successful recovery of complete plantlets. When indole- 3-acetic acid (IAA) was added to the gel matrix (consisting of modified MS), 100% of encapsulated nodal segments of *Dendranthema grandiflora* formed roots (Pinker and Abdel-Rahman, 2005). When silver nitrate ($AgNO_3$) was joined with IBA, the conversion frequency of Chonemorpha grandiflorasyn seeds improved (Nishitha et al., 2006).

*2.5.2.6. Antioxidants*

The antioxidant such as citric acid, ascorbic acid, pyrogallol, phloroglucinol and L-cycteine are used in tissue culture to reduce excessive browning of explants. Ascorbic acid also can increase the number of shoots growing on explant. L-cycteine is used for phenol leaching for explant.

*2.5.2.7. Gelling agents for Synthetic Nutrient Medium*

Based on the consistency, there are mainly two types of plant tissue culture medium, viz. liquid medium used for suspension culture (is being detailed in 2.8) and semisolid or solid medium. Solidified medium is frequently used in plant tissue culture for it easiness to handle. Gels provide a support to tissues growing in stationary conditions. The source of gelling agent in the solid growth medium can also affect results with certain plant species as well as the genotypes within the species. The medium is solidified with a number of solidifying agents- agar, gelrite or phytagel, gelatin, agarose, guar gum, synthetic polymer or starch polymer etc. Each of these is being briefly described below.

**a) Agar**

The most commonly used solidifying agent is agar. It is a polysaccharide obtained from red algae (Gelidium, Gracilaria). Melting point of agar is 100 °C and gelling temperature is 45 °C. Consistency of the culture medium depends on the concentration of agar. At lower concentrations it forms a semisolid medium, whereas at higher concentrations medium become comparatively harder and solid. At very high concentrations of agar the medium become very hard, this may adversely affect the in vitro growth of culture. Hard medium does not allow the diffusion of nutrients into the tissues. Agar is often used in plant tissue culture medium at a concentrations ranging from 0.6-1.0% (w/v). Other forms of agar (agarose, phytagar, flow agar) are also becoming popular for plant tissue culture. Agar has several advantages over other gelling agents:

- Comparatively cheaper than the other gelling agents.
- Agar does not react with the other constituents of the medium.
- In its pure form, it does not contain any nutrients. Thus nutritional studies of in vitro culture will not be affected by the agar.
- It cannot be digested by any growth regulator used in plant tissue culture.
- It is stable in all feasible inoculation and incubation (culture room) temperatures.

**b) Agarose**

Agarose is much more expensive than agar. It is highly purified form of agar prepared from seaweed Gelidium sp. Generally it is used in the synthetic medium at a concentration of 0.4% (w/v).

**c) Gelrite**

Gelrite or phytagel is comparatively cheaper than the agarose and it can be used as an alternative to agar. It is a naturally derived polymer produced by the microbial fermentation of a bacterium Pseudomonas elodae. Only 0.1-0.2% (w/v) of gelrite is required for solidification of the medium. It forms remarkably clear gel than agar and can be used for the study root morphogenesis.

**d) Guar gum**

Guar gum, a galactomannan derived from the endosperms of *Cyamopsistetra gonoloba*, has been successfully used as a sole gelling agent for plant tissue culture media. The gum is commercially extracted from the seeds which are essentially derived by a mechanical process of roasting, differential attrition, sieving and polishing. Its suitability as a gelling agent was demonstrated by using guar gum-gelled media for in vitro seed germination of *Linumusitatissimum* and *Brassica juncea*, in vitro axillary shoot proliferation in nodal explants of *Crataeva nurvala*, rooting of regenerated shoots of the same, in vitro and cogenesis in anther cultures of *Nicotiana tabacum*, and somatic embryogenesis in callus cultures of *Calliandra tweedii*. The media used for these were gelled with either guar gum (2, 3, or 4%) or agar (0.9%). Guar gum-gelled media, like agar media, supported all these morphogenic responses. Rather, axillary shoot proliferation, rhizogenic and embryogenic responses were better on guar gum-gelled media than on agar media. Gaur gum has the following properties:

- It is soluble in hot and cold water but insoluble in most of the organic solvents.
- It has strong hydrogen binding properties.
- It has excellent thickening, emulsion, stabilizing and film forming properties.
- The viscosity of gaur gum is influenced by temperature, pH, presence of salts and other solids.

**e) Carragreenan**

Carragreenan is a textural ingredient with extremely effective gelling properties. It is produced from red seaweed, which can be found along coastlines all over the world. It is named after the Irish town of Carragreenan.

**f) Locust bean gum**

Locust bean gum is also a textural ingredient derived from the seeds of the leguminous carob tree (*Ceratonia siliqua*), which grows in Mediterranean countries. The carob seed consists of three different parts, namely the husk surrounding the seed, the germ (protein) and endosperm (gum). Endosperm is used to extract gum.

*2.5.2.7. pH of medium*

The negative logarithm of the hydrogen ion concentration of any solution for measuring acidity or alkalinity is known as pH. It is a feature of the medium but not a component of the medium. Growth response of plants is pH specific. So, optimum pH of the medium is maintained to facilitate proper function of the cell membrane and uptake nutrients from the medium. pH of plant tissue culture medium is adjusted between 5.0 and 6.0 before adding gelling agent and adjusted with the help of dilute

NaOH, KON or HCl. The pH of an autoclaved media is lowered by 0.3-0.5 unit. In some cases low pH is required for the activation of particular element. For example, Al impairs toxicity in soils having low pH (< 5.0). Free Al-ions are solublized at low pH. Hence, for in vitro screening for Al-toxicity tolerance, the medium pH is usually maintained between 3.8 and 4.0. At higher pH medium become harder, while low pH, the medium will not gel properly.

## 2.6. PREPARATION OF STOCKS SOLUTIONS

For every experiment change in the quantity and quality of media constituents becomes necessary. Plant tissue culture chemicals used for preparing media should be of analytical grade. It is desirable to weigh and dissolve each ingredient separately before mixing them together. Another convenient procedure is to prepare stock solutions, which are mixed together in appropriate quantities, constitute a basal medium. Make up the volume to near the required volume and pH is adjusted. The medium is then brought to its precise volume. Mistakes in the preparation of media can do greater harm than any other fault in the tissue culture technique. It is absolutely essential that all steps in the media preparation and composition should be followed carefully.

Ready-made kits of almost all culture media are also available. These are prepared as per the directions available with the kits. No need to prepare stock solutions. Preparation of basal medium using kits is very simple, saves time and reduces the cost of medium preparation. Other treatments can be given as per requirements of the experiments or commercial purpose.

### 2.6.1. Stock Solution for MS Basal Medium

**Step I.**

The convenient method of media preparation is to prepare the stocks for each basal medium separately, for example in MS medium, major salts are prepared at the concentration of 20×, minor salts, iron, and organic nutrients are prepared at the concentration of 200×. Weigh all the salts as given in the Table 2.5. Prepare the iron stock in coloured (amber) bottle.

**Step II.**

Prepare the growth regulator stock separately usually at the rate of 1 mg/ml (or 1 mmol $L^{-1}$) of stock solution. Growth regulators are prepared by dissolving it in a small quantity of appropriate solvent as given in the Table 2.6 and then adjusted with distilled water to the desired volume.

Table 2.5. Nutritional composition of commonly used nutrient media

| Component | Amount (mg $L^{-1}$) | | | | | |
|---|---|---|---|---|---|---|
| | MS[1] | N6[2] | B5[3] | Nitch's[4] | White's[5] | Gautheret[6] |
| MACRONUTRIENTS | | | | | | |
| $MgSO_4.7H_2O$ | 370.000 | 185.000 | 250.000 | 185.000 | 750.000 | 125.000 |
| $KH_2PO_4$ | 170.000 | 400.000 | - | 68.000 | - | 125.000 |
| $NaH_2PO_4.H_2O$ | - | - | 150.000 | - | 19.000 | - |
| $KNO_3$ | 1900.000 | 2830.000 | 2500.000 | 950.000 | 80.000 | 125.000 |
| $NH_4NO_3$ | 1650.000 | - | - | 720.000 | - | - |
| $CaCl_2.2H_2O$ | 440.000 | 166.000 | 150.000 | - | - | - |
| $(NH_4)_2.SO_4$ | - | 463.000 | 134.000 | - | - | - |
| $Ca(NO_3)_2.4H_2O$ | - | - | - | - | - | 500.000 |
| $NiSO_4$ | - | - | - | - | - | 0.050 |
| MICRONUTRIENTS | | | | | | |
| $H_3PO_3$ | 6.200 | 1.600 | 3.000 | - | 1.500 | 0.050 |
| $H_2SO_4$ | - | - | - | - | - | 1.000 |
| $MnSO_4.4H_2O$ | 22.300 | 4.400 | - | 25.000 | 5.000 | 3.000 |
| $MnSO_4.H_2O$ | - | 3.300 | 10.000 | - | - | - |
| $ZnSO_4.7H_2O$ | 8.600 | 1.500 | 2.000 | 10.000 | 3.000 | 0.190 |
| $Na_2MoO_4.2H_2O$ | 0.250 | - | 0.250 | 0.250 | - | - |
| $CuSO_4.5H_2O$ | 0.025 | - | 0.025 | 0.025 | 0.010 | 0.500 |
| $CoCl_2.6H_2O$ | 0.025 | - | 0.025 | 0.025 | - | - |
| $BeSO_4$ | - | - | - | - | - | 0.100 |
| KI | 0.830 | 0.800 | 0.750 | - | 0.750 | 0.500 |
| $Ti(SO_4)_3$ | - | - | - | - | - | 0.200 |
| $FeSO_4.7H_2O$ | 27.800 | 27.800 | - | 27.800 | - | 0.050 |
| $Na_2EDTA.2H_2O$ | 37.300 | 37.300 | - | 37.300 | - | - |
| EDTA Na ferric salt | - | - | 43.000 | - | - | - |
| SUCROSE (g) | 30.000 | 50.000 | 20.000 | 20.000 | 20.000 | 30.000 |
| ORGANIC SUPPLEMENTS | | | | | | |
| Vitamins | | | | | | - |
| Thiamine HCl | 0.010 | 1.000 | 10.000 | 0.500 | 0.010 | 0.100 |
| Pyridoxine HCl | 0.010 | 0.500 | 1.000 | 0.500 | 0.010 | 0.100 |
| Nicotinic acid | 0.050 | 0.500 | 1.000 | 5.000 | 0.050 | 0.500 |
| Myoinositol | - | - | 100.000 | 100.000 | - | - |

| OTHERS | | | | | | |
|---|---|---|---|---|---|---|
| Glycine | 3.000 | - | - | 2.000 | 3.000 | 3.000 |
| Folic acid | - | - | - | 0.500 | - | - |
| Biotin | - | - | - | 0.050 | - | - |
| pH | 5.8 | 5.8 | 5.5 | 5.8 | 5.8 | 5.8 |

[1]MS:Murashige and Skoog (1962); [2]N6: Chu (1978); [3]B5: Gamborg et al. (1968); [4]Nitch's: Nitch and Nitch (1969); [5]White's: White (1953); [6]Gautheret: Gautheret (1942)

**Table 2.6.** Solubilizing agents of growth regulators

| Growth regulators | Chemical formula | Molecular weight | Solubility |
|---|---|---|---|
| 2,4-Dichlorophenoxyacetic acid | $C_8H_7O_3Cl$ | 221.0 | Alcohol |
| p-Chlorophenoxyacetic acid | $C_8H_6O_3Cl$ | 186.6 | Alcohol |
| Indol-3 acetic acid | $C_{10}H_9NO_2$ | 175.2 | 1N NaOH |
| Indol-3 butyric acid | $C_{12}H_{13}NO_2$ | 203.2 | 1N NaOH |
| α-naphthaleneacetic acid | $C_{12}H_{10}O_2$ | 186.2 | 1N NaOH |
| β-Naphthoxyacetic acid | $C_{10}H_9NO_3$ | 202.3 | 1N NaOH |
| Adenine | $C_5H_5N_5.3H_2O$ | 189.1 | $H_2O$ |
| Adenine sulphate | $(C_5H_5N_5)_2.H_2SO_4.2H_2O$ | 404.4 | $H_2O$ |
| Benzyl adenine | $C_{12}H_{11}N_5$ | 225.2 | 1N NaOH |
| N-isopentenylamino purine | $C_{10}H_{13}N_5$ | 203.3 | 1N NaOH |
| Kinetin | $C_{10}H_9N_5O$ | 215.2 | 1N NaOH |
| Zeatin | $C_{10}H_{13}N_5O$ | 219.2 | 1N NaOH |
| Gibberellic acid | $C_{19}H_{22}O_6$ | 346.4 | Alcohol |
| Abscisic acid | $C_{15}H_{20}O_4$ | 264.3 | 1N NaOH |
| Folic acid | $C_{19}H_{19}N_7O_6$ | 441.4 | 1N NaOH |
| Colchicine | $C_{22}H_{25}NO_6$ | 399.4 | $H_2O$ |

## Step III.

All the stock solutions are stored in an appropriate plastic or glass bottles. The stock bottles are kept in a refrigerator at 4 °C.

**Table 2.7.** Molecular weight of commonly used salts and other chemical compounds for preparation of synthetic media

| MAJOR SALTS | | MINOR SALTS | |
|---|---|---|---|
| Compounds | Molecular weight | Compounds | Molecular weight |
| $NH_4NO_3$ | 80.09 | $H_3BO_3$ | 61.84 |
| $KNO_3$ | 101.10 | $MnSO_4$. $4H_2O$ | 223.09 |
| $MgSO_4.7H_2O$ | 246.50 | $ZnSO_4.7H_2O$ | 287.55 |
| $KH_2PO_4$ | 136.09 | KI | 166.01 |
| $CaCl_2.2H_2O$ | 147.02 | KCl | 74.56 |
| $Ca(NO_3)_2.4H_2O$ | 236.15 | $Na_2MoO_4.2H_2O$ | 241.95 |
| $NH_4Cl$ | 53.49 | $CuSO_4.5H_2O$ | 249.68 |
| $(NH_4)_2SO_4$ | 132.14 | $CoCl_2.6H_2O$ | 237.93 |
| $K_2HPO_4$ | 174.18 | $Na.EDTA.2H_2O$ | 372.20 |
| | | $FeSO_4.7H_2O$ | 278.00 |
| VITAMINS AND AMINO ACIDS | | PLANT GROWTH REGULATORS | |
| Compounds | Molecular weight | Compounds | Molecular weight |
| Biotin | 244.32 | Adenine | 135.13 |
| Calcium pentothenate | 476.53 | Adenine sulphate | 368.34 |
| Folic acid | 441.41 | 2,4-D | 221.04 |
| Penthenic acid | 219.24 | NAA | 186.21 |
| p-Aminobezoic acid | 137.13 | IBA | 203.24 |
| L-Ascorbic acid | 176.12 | IAA | 175.19 |
| Nicotinic acid | 123.11 | Kinetin | 215.22 |
| Riboflavin | 376.36 | 2-ip | 203.25 |
| Thiamin hydrochloride | 337.28 | BAP | 225.26 |
| Pyridoxine hydrochloride | 205.64 | TDZ | 220.25 |
| Glycine | 75.07 | | |
| Myoinositol | 180.16 | | |

Individual nutrient salt can be prepared in a separate bottle or four stocks bottles may be used for preparation of basal MS medium. The molecular weight of major media components are presented in Table 2.7. Stock Solution 1 may contain all the macronutrients (namely $MgSO_4.H_2O$, $KH_2PO_4$, $KNO_3$, $NH_4NO_3$, $CaCl_2.2H_2O$) at appropriate concentrations. Stock Solution 2 should have all the micronutrients ($H_3BO_4$, $MnSO_4.4H_2O$, $ZnSO_4.7H_2O$, $Na_2MoO_4.2H_2O$, $CuSO_4.5H_2O$, $CoCl_2.6H_2O$). Stock Solution 3 contains $FeSO_4.7H_2O$ and $Na_2$.EDTA.$2H_2O$. Finally the Stock Solution 4 contains inositol, thiamine HCl, Pyridoxine HCl, Nicotinic acid and Glycine.

## 2.7. INOCULATION OF EXPLANT

Tissues and cells cultured on an agar gel-medium form an unorganized mass of cells called ***callus***. The synthetic medium is fortified with auxin, namely 2, 4-D, 2, 4, 5-T, TDZ and NAA for callus induction. Sometimes, equal concentrations of auxin and cytokinin also results in callus induction. Continuous dark is required for callus induction; therefore bottom two selves of a rack at corner of culture room should be covered with thick black cloth/curtain. Callus cultures need to be sub-cultured every 3-5 weeks because nutrients depletion takes place in the course of cell growth and medium drying. Production of plantlets through callus culture and maintain true-to-type (genetically identical) will only called as micropropagation, if there is somaclonal variation, it can not be considered as micropropagation.

### 2.7.1. Explant Selection

A wide variation of explants is available for induction of callus. It depends on the species as well as the genotype of plants. The tissue or plant part separated from plant for culture is termed as ***explant*** and the plant from which it is separated is known as ***stock plant***. The establishment of the culture for micropropagation or callus induction also depends on the plant part, the age of the plant, growing environment of the stock plant and finally the composition of the culture medium.

The commonly used explants for callus induction are leaves, stalk, inter node, portion of young twigs, plumule region of germinated seeds, root, rachila, immature inflorescence, anther, microspore, coleoptiles, ovary, ovules, matured seeds, immature seeds, seed embryo, immature embryo etc. (Fig. 2.5). Young and healthy plant parts are used to establish callus. Any meristematic tissues or matured cells are rarely used for raising callus.

### 2.7.2. Explant Preparation

The widely used methods of micropropagation are the axillary, apical shoot meristems and adventitious shoots. Another method is generation of plantlets from callus and the cells of the callus should be genetically identical and true-to-type. In this exercise, bulbs/cloves of tuberose will be used as explant.

### 2.7.3. Surfactant

Tween 20, Triton X-100 and Teepol are scientific reagent-grade surfactants. These are often added in low concentrations (usually 0.05%) to chemical sterilizing solution. Their use ensure that the sterilizing agent comes in contact with the entire plant tissue surface.

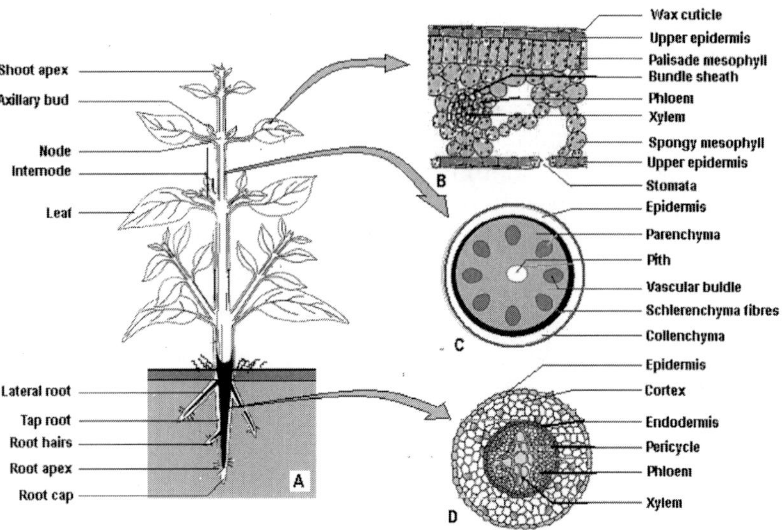


**Fig. 2.5.** Different organs and tissues system in plant body, which can be used as explant for tissue culture. **A)** Whole plant showing different parts, **B)** Cross-section of leaf, **C)** Cross-section of stem, **D)** Cross-section of root.

## 2.7.4. Sterilization of Explant

- Prepare the explant for sterilization.
- Prepare 0.2% $HgCl_2$ solution.
- If there are chances of contaminations of the harvested explant, dip the explant in solution of fungicide (Bavistin).
- Stirring of the tissues: Good surface contact is facilitated by stirring the explant during sterilization.
- Wash the explant in distilled water to remove the fungicide.
- Wash the explant in mild detergent (Teepol).
- Wash the explant 3-4 times in distilled water to remove detergent from the surface of explant.
- Dip the explant in 0.2% $HgCl_2$ solution for 10-15 minutes.
- Decant the $HgCl_2$.
- Wash the explant 3-4 times in sterile distilled water.
- The explants are now ready for inoculation under aseptic environment

Ultrasonic bath: It is an effective method to ensure good surface contact during sterilization treatment in an ultrasonic bath. This technique is particularly useful for sterilization of buds and woody tissues that have many small surface crevices and cracks.

### 2.7.5. Inoculation of Explant for Callus Induction

- Take out the surface sterilized explant and place them on sterile (autoclaved) tissue paper to soak the excess water from the surface of the explant.
- Using sterile forceps, place the explant on autoclaved medium in culture tubes or Petri-dishes or culture bottles for callus induction. Gently press the explants on the medium to fix properly.
- Close the mouth of culture bottles or culture tubes.
- All the activities are done under Laminer Airflow Cabinet and within the periphery of 15 cm diameter around the burner.

Inoculated explants are kept in Culture Room under complete dark and 25 ± 2 °C temperature for callus induction.

## 2.8. SUSPENSION CULTURE

Suspension culture is used to culture isolated single cell in liquid medium. Haberlandt (1902) made first attempt to culture isolated single cell from leaves of flowering plants, but he failed to achieve division of single cell in liquid medium. Continuous progresses in plant tissue culture lead not only to achieve division of single cell in synthetic liquid culture medium, but also now it is possible to regenerate whole plant from single cell. A suspension culture capable of producing a high quantity of proembryogenic masses was evaluated to provide adequate support for synthetic seed production in many crop plants (Utomo et al., 2008).

### 2.8.1. Isolation of Single Cells

#### *2.8.1.1. From plant parts*

Suspension culture can be initiated from any part of plant. The most desirable part to initiate cell culture is leaves. Suspension culture can also be started from sterile seedlings or imbibed seed embryos.

#### *2.8.1.2. From in vitro Grown Callus*

It is most common to initiate suspension culture from in vitro grown callus. Repeated sub-culture of callus on ***callus maintenance medium*** improves the friability of the callus, a pre-requisite for raising a fine cell suspension culture. Few pieces of undifferentiated friable callus are transferred to liquid medium on a gyratory shaker. The nutrient and hormonal composition of the liquid medium is the same as that of ***callus induction medium*** except addition of osmoprotectants and other few changes as per the requirement of the suspension culture. The culture is continuously agitated, usually at a speed of 100-150 rpm. The purposes of continuous agitation of the culture are:

- The gyratory movement of the culture medium exerts pressure on the pieces of callus leading breakage of the callus into free cells and very small cell aggregates.
- It facilitates good gas exchange for cellular respiration of the cells in liquid medium.
- It maintains uniform distribution of cells and small cell aggregates in the liquid medium.

### 2.8.2. Methods of Isolation of Single Cell

Single cell for suspension culture can be isolated either using mechanical or enzymatic methods, which are being briefly discussed below.

#### *2.8.2.1. Mechanical method*

The plant materials (leaves, seedlings or imbibed embryos) gently macerated with a mortar and pestle in presence of a standard grinding solution (20 µmol sucrose, 10 µmol $MgCl_2$, 20 µmoltris-HCl buffer, pH 7.8). The macerated plant parts may contain intact living cells, dead cells and debris of busted cells. The homogenate can be passed through a muslin cloth to separate debris and small pieces of broken plant parts. The screened cells are washed and purified through centrifugation at low speed. The isolated cells are then cultured in liquid medium on an orbital-platform shaker.

#### *2.8.2.2. Enzymatic Method*

The enzymatic method of single cell isolation is convenient over mechanical isolation. This method produces a large number of metabolically active cells. In this method, cell wall is degraded using pectinase to release single cells from plant tissue or culture callus. Enzymatic digestion followed by filtration and centrifugation which leads to the production of free cells which are suitable for suspension culture.

### 2.8.3. Growth Pattern under Suspension Culture

The rate of cell division is much higher in suspension culture as compared to solid and semi-solid cultures. The biomass growth in suspension culture under controlled condition of light, temperature and aeration follows a fixed pattern which is a typical growth curve of 'S' shaped (Fig. 2.6). The growth can be divided into four phases as follow:

#### *2.8.3.1. Lag phase*

The cell growth in this phase is very slow, because the cells adjust to the replenished supply of nutrients and undertake all the necessary synthesis prior to cell division.

#### *2.8.3.2. Exponential phase*

The lag phase is followed by the logarithmic phase or exponential phase of growth. It is the most active phase of cell division. The cell division in this phase is very fast causing logarithmic increase in cell number. Under optimum conditions the cell numbers double every 20-50 hours of culture depending upon the plant species.

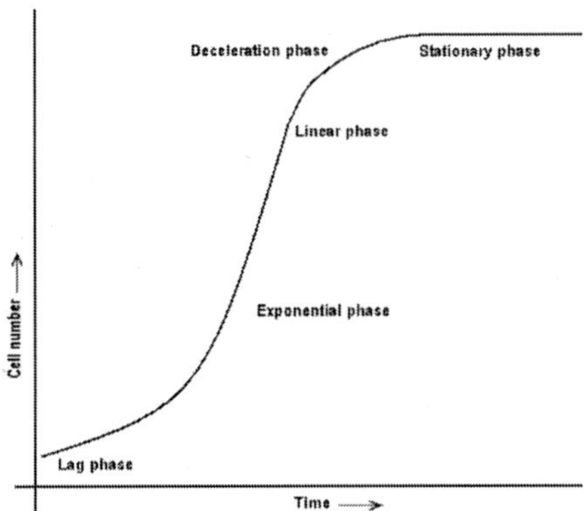


**Fig. 2.6.** Growth curve showing different phases of growth of cells under suspension culture.

#### *2.9.3.3. Linear Phase*

The culture passes through a further phase of rapid cell division. The increase in biomass in this phase is due to the increase in rate of cells expansion and elongation. The rate of cell division slows down at the end of this phase as the nutrients in the culture medium get depleted.

#### *2.8.3.4. Deceleration Phase*

After 3-4 cell generations the cell division and growth ability declines. Finally the cell population enters into the stationary phase.

#### *2.8.3.5. Stationary Phase*

The cell division and the growth of individual cell cease, thus the cell number and cell mass remains static in this phase. The nutrients in the culture medium are almost depleted, some cells enter senescence phase. If the culture medium is not changed and cells are allowed to remain in the old medium, the cells may die by the end of stationary phase.

### 2.8.4. Maintenance of Cell Suspension Culture

After exponential phase of growth, if the culture is left for longer time, the cells may die. If the culture is to be maintained further, a portion of culture at linear growth phase, particularly the single cells and small cell clumps is transferred to a fresh culture medium. The cells are collected using a pipette with a fine orifice or using sieve of appropriate size and transferred to the fresh subculture medium.

### 2.8.5. Types of Suspension Culture

#### *2.8.5.1. Batch Culture*

The volume culture medium of batch culture remains fixed. By the end exponential growth phase, the nutrients in the batch culture medium gets depleted leading to the cease of cell division and there will be further increase in cell mass. The culture is agitated on a gyratory shaker at a speed of 60-100 rpm for proper growth and division of cells. To maintain the culture, a small volume of cultured cells is transferred to a new vessel containing fresh suspension culture medium. Batch culture is characterized by constant change in the pattern of cell growth and metabolism. The main draw-back of batch culture is no period of steady-state. To overcome this draw-back, continuous culture system may be practiced.

#### *2.8.5.2. Continuous Culture*

In this system of culture, a steady-state of cell growth is maintained for long period by addition of fresh medium in the culture vessel simultaneously draining out /harvesting the culture so that the culture is maintained indefinitely. This method is used for production of cells in large-scale to extract secondary metabolite.

### 2.8.6. Use of Suspension Culture

- To develop ***cell line***. It is possible to raise single cell from suspension cultures and raise a callus from them.
- For production of embryogenic callus.
- For commercial production of cell mass.
- For extraction of secondary metabolite from isolated cell line.
- Suspension culture may be used for study of effect of different chemicals on cell culture, cell development and cell division etc.

## 2.9. PROBLEMS ASSOCIATED WITH PLANT TISSUE CULTURE AND THEIR POSSIBLE SOLUTIONS

The main objective of plant tissue culture is to develop pathogen free good culture followed by production of complete plant with well developed roots and shoot. However, a number of problems are associated with the aseptic culture. The common problems, causes and their possible solutions are being listed in Table 2.8.

**Table 2.8.** The probable problems associated with plant tissue culture and their possible remedy (modified from Bhansali, 1995)

| Symptoms | Probable causes | Possible solution |
|---|---|---|
| Culture contamination | Infested with fungi or bacteria<br>Inappropriate sterilization | Follow appropriate sterilization procedure and use appropriate surface sterilizing agent |
| | Systemic infection | Use antibiotics for systemic infection |
| | Inappropriate method of inoculation | Maintain aseptic conditions during inoculation |
| Watery and sticky callus | Bacterial contamination | Use antibiotic in the medium |
| Fungal contamination on medium, not the culture | Contaminated medium | Transfer the culture to fresh medium |
| Medium discolourization | Release of phenolic | Use activated charcoal (0.5-3.0%) in the medium<br>Use polyvinylpyrrolidone (250-1000 mg $L^{-1}$ |
| Greening of callus before embryogenesis | High light intensity | Keep the callus in dark till callus is transferred to regeneration medium |

| | | |
|---|---|---|
| Explant dies | Use of high concentration of sterilizing agent | Use optimum dose of sterilizing agent or change the sterilizing agent |
| | Medium too strong | Reduce the strength of the medium |
| | Wrong stage of plant growth | Obtain explant from different stages of growth and standardize the appropriate growth stage |
| Culture blackens and dies | Contaminated by microorganism(s) | Discard the infected portion followed by ***disinfection*** with mild sterilizing agent |
| | Bleeding | Subculture immediately and frequently |
| | Agar problem or water problem or wrong formulation of media constituents | Check water purity, agar concentration, and formula of media constituent |
| No growth of explant | Dormant | Chill for a month<br>Take explant when plant tissue are active |
| | Medium too harsh | Use explant at different stages of growth |
| | Wrong formulation of medium | Use different formulation or strength of medium with hormonal concentrations and combinations |
| No growth of culture | Culture room too cold | Maintain optimum temperature |
| | Wrong formulation | Try different formulation or strength of medium with hormonal concentrations and combinations |
| Lanky growth of culture, poor multiplication, yellow and watery leaves | Very low concentration of cytokinin | Use optimum concentration of cytokinin<br>Try for proper ratio of auxin and cytokinin |
| Less vertical growth with leafy shoot | High concentration of hormones | Optimize the concentration of hormone or combination of hormones |
| No multiplication | Low concentration of cytokinin | Optimize the concentration of hormone or combination of hormones |
| | Cold pretreatment needed | Store at ~4 $^0$C for 4-8 weeks |
| | Dormant explant | Take explant at active growth stage of the plant<br>Give cold treatment for 3-8 weeks |
| Only root growth (rhizogenesis) | Improper hormone or hormonal combination | Use proper combination of hormone |
| Only shoot growth (caulogenesis) | Hormonal imbalance | Transfer the shoot to rooting medium supplemented with IAA/ IBA |

| | | |
|---|---|---|
| Insufficient number of embryos | Rupture of intercellular connections and isolation of cells that are reinstated into the condition of zygote | Suspend the cells in a plasmolizing solution containing high concentration of sucrose (1M) for a short period and replaced on normal condition |
| Pale and small leaves | High concentration of hormones (particularly cytokinin) | Use optimum concentration of cytokinin/auxin |
| Chlorotic leaves | Contaminated with pathogen | Index for contamination and use decontaminant |
| | Culture room too hot | Decrease temperature |
| | Wrong medium formulation | Try different media |
| Succulent leaves and abnormal stem | Improper osmotic potential | Decrease temperature |
| | High concentration of cytokinin | Decrease cytokinin concentration<br>Optimize the hormonal combination |
| | Culture too old | Subculture at appropriate interval |
| | Wrong agar | Try different agar as well as different concentrations |
| Premature rooting | Improper hormonal combination | Run cytokinin/auxin grid<br>Increase cytokinin and decrease auxin concentration |
| Red stem/ embryos/ cells | Stress | Adjust light and/ or temperature<br>Decrease sucrose concentration in the medium<br>Increase nitrate ($NO_3$)<br>Subculture more frequently |
| Non-friability of callus | High concentration of sugar<br>Not enough $NO_3$<br>Improper medium composition | Try different media with different hormonal combinations<br>Select most actively growing callus |
| | Culture too old | Subculture more frequently |
| | Plant source | Try different explant |
| Formation of mature embryos, no further growth | Hormonal imbalance | Reduce the hormone concentration in the medium<br>Transfer to cytokinin rich (particularly kinetin and BAP) medium |
| Vitrification-water-soaked, translucent and glassy morphological appearance, particularly leaves and stem | Tightly closed culture vessel<br>Increase level of $CO_2$ in culture vessel<br>High level of ethylene and water vapour | Use cotton-plug to facilitate air diffusion<br>Losses the cap of the culture vessels |

## References


Arun Kumar MB, Vakeswaran V, Krishnasamy V. 2005. Enhancement of synthetic seed conversion to seedlings in hybrid rice. Plant Cell Tiss Org Cult. 81: 97-100.

Bhansali R. 1995. A Manual of Plant Tissue Culture. Department of Biotechnology, Ministry of Science and Technology, Government of India, New Delhi.

Chand S, Singh AK. 2004. Plant regeneration from encapsulated nodal segments of Dalbergiasissoo Roxb. -a timber yielding leguminous tree. J Plant Physiol. 161: 237-243.

Chu CC. 1978.The $N_6$ medium and its applications to anther culture of cereal crops. In: Proceedings of Symposium on Plant Tissue Culture. Science Press, Perking. pp. 43-50.

Gamborg OL, Miller RA, Ojima K. 1968. Nutrient requirement of suspension cultures of soybean root cells. Exp Cell Res. 50: 151-158.

Gautheret RJ. 1942. Manual Technique de Culture de tissues Végétaux. Masson Publ, Paris.

Germanà MA, Micheli M, Chiancone B, Macaluso L, Standardi A. 2011. Organogenesis and encapsulation ofin vitro-derived propagules of Carrizo citrange [Citrus sinesis(L.)Osb. × Poncirius trifoliate (L.) Raf.]. Plant Cell Tiss Org Cult. 106: 299-307.

Haberlandt G. 1902. Kulturversuchemitisolliertenpflanzenzellen.Gizungsber.Akad.Wiss.Wien., Math.-Naturwiss. KI, Abt. 1. 111: 69-92.

Hung CD, Trueman SJ. 2012. Preservation of encapsulated shoot tips and nodes of the tropical hardwoods Corymbiatorelliana× C. citriodoraand Khayasenegalensis. Plant Cell Tiss Org Cult. 109: 341- 352.

IUPAC (International Union of Pure and Applied Chemistry). 1980. Atomic Weights of the Elements 1979. Pure Appl Chem. 52: 2349–84.

Linsmaier EM, Skoog F. 1965. Organic growth factor requirements of tobacco tissue cultures. Physiol Plant. 18: 100-127.

Murashige T, Skoog F. 1962. A revised medium for rapid growth and bioassays with tobacco tissue cultures.Physiol Plant. 15: 473-497.

Nishitha IK, Martin KP, Ligimol, Beegum AS, Madhusoodanan PV. 2006. Microproapagation and encapsulation of medicinally important Chonemorphagrandiflora. In Vitro Cell. DevBiol.Plant. 42: 385-388.

Nitsch JP, Nitsch C. 1969. Haploid plant from pollen grains.Science. 163: 85-87.

Pattnaik SK, Sahoo Y, Chand PK. 1995. Efficient plant retrieval from alginate encapsulated vegetative buds of mature mulberry trees. Sci Hort. 61: 227-239.

Piccioni E, Capuano G, Standardi A. 1997: Conversion of M.26 encapsulated microcuttings on perlite. In Vitro Cellular and Developmental Biology. 33(3-II): 51 A.

Pierik RLM. 1989. In Vitro Culture of Higher Plants.MartinusNijhoff Publishers, Dordrecht, the Netherlands. pp. 1-344.

Pinker I, Abdel-Rahman SSA. 2005. Artificial seed for propagation of Dendranthema× grandiflora(Ramat.). PropagOrnan Plant. 5: 186-191.

Roy B, Mandal AB. 2004. Toward development of mapping population through anther culture and conventional recombination breeding for moleculer tagging of salt-tolerant gene/s involving IR 28 and Pokkali. In: 9th National Rice Biotechnology Network Meeting, New Delhi, India, pp. 183-185.

Roy B, Mandal AB. 2005. Anther culture response in indica rice and variation in major agronomic characters among the androclones of a scented cultivar, Karnal local.African J Biotechnol. 4(3): 235-240.

Saiprasad GVS. 2001. Artificial seeds and their applications. Resonance. 2001: 39-47.

Soneji JR, Rao PS, Mhatre M. 2002. Germination of synthetic seeds of pineapple (Ananascomosus L. Merr.). Plant Cell Rep. 20: 891-894.

Swamy MK, Balasubramanya S, Anuradha M. 2009. Germplasm conservation of patchouli (Pogostemoncablin Benth.) by encapsulation of in vitro derived nodal segments. Int J Biol Con.1: 224-230.

Utomo HS, Wenefrida I, Meche MM, Nash JL. 2008. Synthetic seed as a potential direct delivery system of mass produced somatic embryos in the coastal marsh plant smooth cordgrass (Spartinaalterniflora). Plant Cell Tiss Organ Cult. 92(3): 281-291.

Wang WG, Wang SH, Wu XA, Jin XY, Chen F. 2007.High frequency plantlet regeneration from callus andartificial seed production of rock plant Pogonatherumpaniceum (Lam.) Hack. (Poacaecae). Sci Hort. 113: 196-201.

White PR. 1954. The Cultivation of Animal and Plant Cells. 1st. edition. Ronald Press, New York.

White PR. 1963. The Cultivation of Animal and Plant Cells ($2^{nd}$ed.). 2 Ronald Press, New York. pp. 1-239.

# 3

# Gelling Agents and Additives

To mimic the natural seeds, the embryos from cultures are encapsulated in a nutrient gel containing essential organic/inorganic salts, carbon sources, plant hormones and antimicrobial agents and coated completely to protect the embryos from mechanical damage during handling and to allow the development and germination without any undesirable variation. When looking at the history of synthetic seed technology, a wide number of encapsulating agents have been tested in time for their capacity to produce beads like, potassium alginate, sodium alginate, carregeenan, agar, agarose, guargum, polyox, gelrite, sodium pectate, carboxymethyl cellulose, nitrocellulose, ethylocellulose, polyethyleneamine (Kersulec *et al*., 1993), chitosane (Tay *et al*., 1993), tracanth gum etc. have been tested as hydrogels (Table 3.1), out of which alginate emasculation was found to be more suitable and practical for synthetic seed production. Some plants exudates of arabic (*Acacia senegal*), karaya (*Stereculsia* sp.), or seed gums of gaur (*Cyamopsis tetragonoloba*), tamarind (*Tamarindus indica*) or microbial products like dextran, xantham are also being used for encapsulation.

**Table 3.1.** Gelling materials and complexing agent for encapsulation of plant propagules for production of synthetic seeds

| Gelling agent | | Complexing material | |
|---|---|---|---|
| Substance | Concentration (%) | Substance | Concentration (mM) |
| Sodium alginate | 0.5-5.0 | Calcium salts | 30-100 |
| Sodium alginate with gelatin | 2.0 | Calcium chloride | 30-100 |
| Carrageenan | 0.2-0.8 | Potassium chloride | 500 |
| Locust Beam Gum | 0.4-1.0 | Ammonium chloride | 500 |
| Gelrite* | 0.25 | - | - |
| Agar* | 0.8-2.0 | - | - |

*Temperature is lowered down for solidification

## 3.1. GELLING AGENTS

### 3.1.1. Polyoxyethylene oxide

Kitto and Janick (1985a) tested eight chemical compounds for the production of synthetic seed coats, among which a water-soluble resin, the polyethylene oxide homopolymer (polyox), was the best to coat embryogenic suspensions (i.e., cells, aggregates, callus clumps and embryos) of carrot.Later on, this methodology was used to obtain synthetic seeds of orange and celery as well (Kitto and Janick 1985b, Kim and Janick 1989). Polyoxyethylene oxide is readily soluble in water, dries to form a thin film, does not support the growth of microorganisms and is non-toxic to the embryo thereby leading to the production of desiccated synthetic seed. Desiccation can be achieved either slowly over a period of one or two weeks sequentially using chambers of decreasing relatively humidity, or rapidly by unsealing the Petri dishes and leaving them on the bench overnight to dry.Polyox was reported to have several positive characteristics as an encapsulating agent, such as-

1. It formed a film after drying,
2. It didn't support the growth of contaminants, and
3. It was non-toxic for the encapsulated embryos.

### 3.1.2. Alginate Hydrogel

Alginate is a straight chain, hydrophilic, colloidal polyuronic acid composed primarily of hydro-β-D-mannuronic acid residues with 1-4 linkages. Redenbaugh *et al.* (1988), after comparing several substances for making synthetic seed coats, proposed the use of a Na-alginate solution which could be turned into a hardened Ca-alginate gel by an ion-exchange reaction. Somatic embryos of alfalfa and celery were the first to be alginate-coated, reaching an artificial seed germination rate, which in alfalfa was over 85%. Since then, the alginate became by far the most used gelling agent for synthetic seed preparation.

Sodium alginate has a low cost and good gelatin and biocompatibility characteristics for preparation of synthetic seeds. The major principle involved in the alginate encapsulation process is that the sodium alginate droplets containing the somatic embryos when dropped into the $CaCl_2.2H_2O$ solution form firm bead due to ion exchange between the $Na^+$ in sodium alginate with $Ca^+$ in the $CaCl_2.2H_2O$ solution. The hardness of beads depends on the number of sodium ions exchanged with calcium ions. Generally 3% sodium alginate and 75 mM $CaCl_2.2H_2O$ for half an hour for optimum rigidity of encapsulated beads. Hence, the rigidity of the alginate capsule provided better protection to the encased plant materials. At the same time, the internal factor related to microshoots could also be one of the important factors for limiting germination.

The frequency of conversion from encapsulated embryos to plantlets was significantly affected by the concentration of sodium alginate, concentration and source of complesing agent. The concentration of sodium alginate may differ depending on the requirement and rigidity/ tolerance ability of callus or other explants. Arun Kumar *et al.* (2005), Nair and Gupta (2007) used as high as 4% sodium alginate and 1.5% calcium chloride solution for preparation of synthetic seeds of hybrid rice and as low as 0.8% sodium alginate (Patel *et al.*, 2000) for encapsulation of shoot tips of *Solanum tuberosum*. However, increased concentration of sodium alginate may affect the germination of somatic embryos by affecting the degree of vigour or maturity of the embryos at the moment of being capsulated, given that the mechanical resistance that an excessively hard encapsulation can result in a major part of the available energy is being utilized in breaking the synthetic endosperm. It was observed that a concentration $\geq$ 5.0% (v/w) drastically reduce the percentage of germination of encapsulated somatic embryos Nair and Gupta (2007).

Alginate hydrogel is frequently used as a matrix for synthetic seed preparation because of its following characteristics:

- The excellent water solubility
- Moderate viscosity of Na-alginate at room temperature
- Low spin-ability of solution
- Low toxicity for somatic embryos or absence of any kind of toxicity of the Ca-alginate matrix for explants
- Long-term storability of the Na-alginate solution
- The easy use of calcium salts for quick gellation and bead hardening at room temperature
- Quick solidification
- The possibility to prepare artificial seeds of different hardness by changing the concentration of Na-alginate and/or the duration of the ion-exchange reaction
- Possibility to mix the alginate with a nutritive medium to obtain an artificial endosperm

- Bio-compatible
- The rigidity of alginate beads provides an effective protection to the somatic embryos
- It is easily available at low cost

Though it has many advantages as gelling agent for preparation of artificial seeds, it also carries some unfavourable characteristic, such as-

- The tacky nature of the capsules that are produced
- Indeed, the alginate seed coats are very moist and they have a sticky surface, making the seeds adhere to each other and difficult to separate
- Another limitation is due to the rapid dehydration of the alginate beads which, upon exposure to air, makes the artificial seeds become very hard in just a few hours, preventing or making the conversion of the enclosed explants into plantlets or shoots very difficult.

#### 3.1.3. Agar as Gelling Agent

The use of agar as gelling agent for synthetic seed production is avoided as it is considered inferior to alginate beads. Moreover, agar remains in solution form only comparative at high temperature, which may be injurious to somatic embryos. Finally bead preparation using agar is difficult. *In vivo* sowing synthetic seed prepared by agar is almost impossible, as it is highly susceptible to fungal infection.

### 3.2. COMPLEXING MATERIALS

#### 3.2.1. Calcium Chloride

Calcium chloride is used at a concentration ranging from 30-100 mM. $Na^+$ ion in sodium alginate and is replaced by the $Ca^+$ ion in the $CaCl_2.2H_2O$ solution leading to solidification subsequently in which this reaction from the beads. High concentration or excessive exposure to $CaCl_2.2H_2O$ solution may results in more absorption and penetrations of $CaCl_2$ to embryo, which may generate growth inhibition that can reflects in decrease in the germination response and subsequent development in the field (Malabadi and Van Staden, 2005).

### 3.3. ADDITIVES TO THE MATRIX

A number of useful materials can be added to the gelling agents to supply nutrients to somatic embryos and to prevent the somatic from mechanical injuries, namely organic and inorganic nutrients, fungicides, insecticides, antibiotics, charcoal and microorganisms. The possible adjuvant used in preparation of synthetic seeds is being briefly discussed below.

### 3.3.1. Activated Charcoal

Addition of activated charcoal improves the conversion and vigour of encapsulated somatic embryos. It has been reported that the charcoal breaks up the alginate and thus increases respiration of somatic embryos. It also retains the nutrients within the hydrated gel and release slowly to the growing embryo. Arun Kumar *et al.* (2005) reported that inclusion of activated charcoal has enhanced the germination and conversion to the maximum extent by increasing the diffusion of gases and nutrients and respiration of embryoids. Activated charcoal (0.1%) can also be added to the matrix to absorb the polyphenol exudates of the encapsulated shoots of banana (Ganapathi *et al.,* 1992).

### 3.3.2. Use of Plant Growth Regulators

To improve the storability and germination of synthetic seed, many workers used growth regulators.

#### *3.3.2.1. Germination and Plant Establishment*

Sucrose is not only an important substance in capsule matrix but also in culture media. The presence of sucrose in germination medium showed significant effect on germination rate and plantlets survival rate of *Oryza sativa* L. Cv. MRQ 74. Similar findings were reported on other species such as *Psidium guajava* L. With increasing concentration of sucrose (3–9%) in medium, the percentage of germination of encapsulated somatic embryos of *Psidium guajava* L. decreased significantly(Rai *et al.*, 2008). Taha *et al.* (2009) found that MS without hormones supplemented with 30 g/L sucrose was the best substrate for germination of the synthetic seeds of *Saintpaulia ionantha* Wendl. The use of tap water in culture medium and substrate (top soil) gave maximum germination

Buds treated with 0.01-1.0 mM ABA either prior to encapsulation or even in the alginate, matrix to inhibit the precocious growth (Palmer and Jasrai, 1996). The ability of somatic embryos to with stand to low moisture content is important for storage, and also plays a big role in developmental transition between maturation and germination. Phokajornyod *et al.* (2004) found that somatic embryos that treated with 0.5 mg/l of ABA for 20 days before encapsulation with sodium alginate, and dehydrated in laminar flow hood until 80% water loss still remained germinated to 58%. Furthermore, improvement of dry somatic embryos was accomplished by adding 60 g/L sucrose in the maturation medium which resulted 40% of plantlets conversion after four weeks of storage in ambient temperature.

Ammirato (1974) showed that ABA at $10^{-7}$ M prevented precocious germination of somatic embryos of caraway (*Carum carvi*) in suspension culture. Later Amirato (1983) reported that the same level of ABA had a similar effect on suspension cultured carrot somatic embryos, producing embryos more similar to their zygotic counterparts than those grown without ABA. Base on those results, he proposed that regeneration of embryo maturation by ABA might be used to facilitate large-scale batch cultures, mechanized planting, artificial induction of dormancy and incorporation into artificial seeds. Roberts *et al.* (1990) found that the presence of ABA essential for estimation of storage protein accumulation in somatic embryos of interior spruce.

#### *3.3.2.2. Storability Improvement*

Encapsulation with $GA_3$ was found to be useful for storage of somatic embryos of citrus at 4 °C for 1 month (Mariani, 1992; Antonietta *et al.*, 1998). Synthetic seeds obtained from the excised embryos of intact seeds treated with higher (2-3 mg/$l^{-1}$) concentrations of ABA showed tolerance to low temperature storage and retained higher germination percentage. ABA is implicated as a controlling factor for germination and dormancy in somatic embryos and seeds (Senaratna *et al.*, 1995). Ruffoni *et al.* (1994) found that the addition of zeatin (0.5 mg/l) to the alginate encapsulation coating improved the shoot production and sucrose (40 g/l) added to MS agar medium improves the percentage root emergence of *Enstoma grandiflorum*.

#### *3.3.2.3. Induction of Desiccation Tolerance*

Desiccation tolerance is a quantitative character and embryos can be of varying degrees of tolerance. This is especially true among different species, but is also evident with different inductive treatment (McKersie*et al.*, 1994). Desiccation tolerance is a characteristic of somatic embryos that must be induced and therefore, requires a pretreatment with ABA or stress elicit the desired response. The importance of ABA application for imparting desiccation tolerance during storage of somatic embryos was well recognized (Senaratna *et al.*, 1990; Lecouteux *et al.*, 1993). Kitto and Janick (1985a) reported that ABA effectively hardened carrot embryos, permitting them to survive desiccation. In a further study, Kim and Janic (1989) found that ABA at $10^{-6}$ M effectively increased embryo survival after desiccation suggesting that ABA hardens somatic embryos of celery. They also achieved desiccation tolerance by a combination of ABA and proline. ABA increased the accumulation and altered the distribution of fatty acids in somatic embryo. Takahata *et al.*, (1992) reported that the desiccated embryos lost their viability if not treated with ABA. Dormancy and desiccation tolerance can be imparted to the embryos by treating them with appropriate concentration of ABA or high sucrose concentrations (Anandarajah and McKerrie, 1990a).

### 3.3.3. Use of Protective Chemicals

Pesticides and fertilizers can be incorporated into capsules to enhance germination rate and seedling growth and protection from disease as well as insect-pests during germination. The composition of the protective matrix should allow the growth of the encapsulated embryo, providing mechanical resistance according to the available energy of the embryo, given that an excessively hard endosperm results in energy loss and weak or no growth of the encapsulated somatic embryo.

The microscopic destroying worms could be packed during encapsulation (Redenbaugh *et al.*, 1986). To avoid bacterial contamination, Ganapathi *et al.* (1992) added an antibiotic mixture (0.25 mg/l) containing rifampicin (60 mg), cefatoxime (250 mg) and tetracycline-HCl (25 mg) dissolved in 5 ml dimethyl sulphate to get matrix, Arun Kumar *et al.* (2005) used streptomycin in the alginate matrix. To avoid contamination of synthetic seeds prepared from protocorm-like bodies of *Geodorumdens iflorum* under *in vivo* condition, Datta *et al.* (1999) used fungicide (bavastin) and food preservative (sodium bicarbonate) which did not show any contamination of synthetic seeds up to eight weeks. Pesticides and fertilizers have been used in preparation of synthetic seeds to enhance germination rates and seedling growth (Toruan and Sumaryono, 1994; Arun Kumar *et al.*, 2005). The incorporation of these preservative and growth enhancing nutrients did not show any effect on germination and conversion percentage.

### 3.3.4. Cyanobacterial Extract

The effect of cyanobacterial extracts (*Plectonema boryanum* UT × 594) was tested for growth and multiplication of shoots from shoot tips as well as encapsulated shoot tips of many crops. Cyanobacterial extracts are known to stimulate somatic embryogenesis and conversion of synthetic seeds (Wake *et al.*, 1992; Bapat *et al.*, 1996).

### 3.3.5. Microorganisms

Zhang *et al.* (2001) prepared synthetic seeds of *Dendrobiumcandidum* using clay and vermiculate powder as the encapsulating medium. Their results showed that the germination rate reached 56.8% when the proportion of clay: vermiculite: water was 2: 1: 2. When this system was added with only 1.0% activated charcoal, or 1.0% activated charcoal along with 0.5% starch, the corresponding average germination rates of the artificial seeds increased to 76.7 and 80.3%. The treatment enhanced the germination rate by 18.4 and 24.2%, respectively. Sharma *et al.* (2000) used mycorrhiza for preparation of synthetic seeds.

## 3.4. GELLING AGENT DISSOLVING MEDIUM

The gelling agents can be dissolved in either distilled water or liquid nutrient medium. Generally it is dissolved in liquid nutrient medium, particularly MS basal medium. The medium can be fortified with plant growth regulator or pesticides as per requirement. The gelling agent in the synthetic seed along with nutrient medium acts as endosperm/ cotyledon of the seed (see Fig. 4.1). In this gelling matrix other additives also can be added to improve the conversion rate.

### References

Ammirato PV. 1974. The effects of abscisi acid on the development of somatic embryos from cells of caraway (*Carumcarvi* L.). Bot Gaz. 135: 328-337.

Ammirato PV. 1983. The regulation of somatic embryo development in plant cell culture: Suspension culture technique and hormone requirements. BioTechnol. 1: 68-74.

Anandarajah K, McKersie BD. 1990a. Manipulating the desiccation tolerance and vigour of dry somatic embryos of *Medicago sativa* L. with sucrose, heat shock and abscisic acid.Plant Cell Rep. 9: 451-455.

Antonietta GM, Emanuele P, Alvaro S. 1998. Effects of encapsulation on *Citrus reticulate* Blanco somatic embryo conversion. Plant Cell Tiss Org. Cult. 55(3): 235-237.

Arun Kumar MB, Vakeswaran V, Krishnasamy V. 2005. Enhancement of synthetic seed conversion to seedlings in hybrid rice. Plant Cell Tiss Org. 81: 97-100.

Bapat VA, Iyer RK, Rao PS. 1996. Effect of cyanobacterial extracts on somatic embryogenesis in tissue cultures of sandalwood (*Santalum album* L.). J Med & Aromatic Plant Sci. 18: 10-14.

Datta KB, Kanjilal B, Sarker D-de, De-Sarkar D. 1999. Artificial seed technology: Development of a protocol in *Geodorumdensiflorum* (Lam) Schltr. – an endangered orchid. Current Sci. 76(8): 1142-1145.

Ganapathi TR, Suprasanna P, Bapat VA, Rao PS. 1992. Propagation of banana through encapsulated shoot tips. Plant Cell Rep. 11: 571-575.

Kersulec A, Baszinet C, Corbineau F, Come D, Barbotin JN *et al.* 1993. Physiological behaviour of encapsulated somatic embryos. Biomater Art Cells ImmobBiotechnol. 21: 375-381.

Kim YH, Janick J. 1989. ABA and polyoxencapsulation or high humidity increase survival of desiccated somatic embryos of celery. Hort Sci. 24: 674-676.

Kitto SL, Janick J.1985a. Production of synthetic seeds by encapsulating asexual embryos of carrot. J American SocHort Sci.110: 277-282.

Kitto SL, Janick J. 1985b. A citrus embryo assay to screen water soluable resins as synthetic seed coats.HortScience.20: 98-100.

Lecouteux C, Lai FM, McKersie BD. 1993. Maturation of somatic embryos by abscisic acid, sucrose and chilling. Plant Sci. 94: 207-213.

Malabadi R, Van Staden J. 2005. Storability and germination of sodium alginate encapsulated somatic embryos derived from the vegetative shoot apices of mature *Pinuspatula* trees. Plant Cell Tiss Organ Cult. 82: 259-265.

Mariani P. 1992.Egg plant somatic embryogenesis combined with synthetic seed technology. Capcium New let. (Special Issue), 7-10 September, 1992. pp. 289-294.

McKersie BD, Van Acker SDN, Lai FM. 1994. Artificial seeds- A comparison of desiccation tolerance in zygotic and somatic embryos. In: Biotechnological Application of Plant Culture, Somatic Embryogenesis and Artificial Seeds, Bajaj YSP (ed.). Springer-Verlag, Berlin. pp. 129-150.

Nair RR, Gupta SD. 2007. *In vitro* plant regeneration from encapsulated somatic embryos of black paper (*Piper nigrum* L.). J Plant Sci. 2(3): 283-292.

Palmer JP, Jasrai YT. 1996. Precocious growth and effect of ABA: encapsulated buds of *Kalanchoetubiflora*. J Plant BiocheBiotechnol. 5(2): 103-104.

Patel AV, Pusch I, Mix-Wagner G, Vorlop KD. 2000. A novel encapsulation technique for the production of artificial seeds. Plant Cell Rep. 19: 868-874.

Phokajornyod P, Pawezik E, Vearsaip S. 2004. Dry synthetic seed production and desiccation tolerance induction in somatic embryos of sweet papper.DeutscherTropentag, Berlin, October, 2004. pp. 5-7.

Rai MK, Jaiswal VS, Jaiswal U. 2008. Effect of ABA and sucrose on germination of encapsulated somatic embryos of guava (*Psidiumguajava* L.) *Scientia Horticulturae.* 117(3):302-305.

Redenbaugh K, Fujii JA, Slade D. 1988. Encapsulated plant embryos. In: Mizrahi A (ed) *Biotechnology in Agriculture- Advances in Biotechnological Processes* (Vol. 19), Alan R LissInc, NY. pp. 225-248.

Redenbaugh K, Paasch BD, Nichol JW, Kossler ME, Viss PR, Walker KA. 1986. Somatic seeds: Encapsulation of asexual plant embryos. Bio/technology. 4:797-801.

Roberts DR, Flinn BS, Webb DT, Webster FB, Satton BCS. 1990. Abscisic acid and indol-3-butyric acid regulation of maturation and accumulation of storage proteins in somatic embryos of interior sprouce. Physiol Plant. 78: 355-360.

Ruffoni B, Massabo F, Giovannini A. 1994. Artificial seed technology in ornamental species *Lisianthus* and *Genista.*Acta Horticulturae. 362: 277-304.

Senaratna T, McKersie BD, Bowley SR. 1990. Induction of desiccation tolerance in somatic embryos.*In vitro* Cellular Dev Biol. 26(1): 85-90.

Senaratna T, Saxena PK, Rao MV, John Afele. 1995. Significance of *Medicago sativa* L. somatic embryos. Plant Cell Rep. 14: 375-379.

Sharma S, Kashyap S, Vasudevan P. 2000. Development of clones and somaclones involving tissue culture, mycorrhiza and synthetic seed technology. J Scientific Industrial Res. 59(7): 531-540.

Taha RM, Daud N, Hasbullah NA, Awal A. 2009. Somatic embryogenesis and production of artificial seeds in *Saintpauliaionantha* Wendl. In: Geijskes RJ, Lakshmanan P, Taji A, editors. Proceedings of the 6th International Symposium on *In Vitro* Culture and Horticultural Breeding. Brisbane, Australia. pp. 331-336.

Takahata Y, Wakui K, Kaizuma N, Brown DCW. 1992. A dry artificial seed system for *Brassica* crops. ActaHorticulturae. 319: 317-322.

Tay LF, Koh LK, Loh CS, Khor E. 1993.Alginatechitosancoacervation in production of artificial seeds. BiotechnolBioeng. 42: 449-454.

ToruanMathius, Sumaryono N. 1994. Application of synthetic seed technology for mass clonal propagation of crop plants.Bulletin Bioteknologi Perkebunan. 1(1): 10-16.

Wake H, Akasaka A, Umetsu H, Ozeki Y, Shimamura K, Matsunaga T. 1992. Enhanced germination of artificial seeds by marine cyanobacterial extract.Appl Micro Biotechnol. 36: 684-688.

Zhang M, Wei Xiao Y, Huang HR, Zhang M, Wei XY, Huang HR. 2001. A study on the solid enasculation system of the artifical seed of *Dendrobiumcandidum*. Acta Horticulturae Sinica. 28(5): 435-439.

# 4

# Artificial Seeds Preparation Technology

The artificial seed production technique was first used in clonal propagation to cultivate somatic embryos placed into an artificial endosperm and constrained by an artificial seed coat. Today artificial seeds represent capsules with a gel envelope, which contains not only somatic embryos but also axillary and apical buds or stem and root segments (Vdovitchenko and Kuzovkina, 2011).

The prime requirements for preparation of artificial seeds are high frequency somatic embryogeniccalli, synchronous embryo development and maximum conversion of embryos into plants. Success in production of synthetic seeds mainly depends on how best callus development and plantlet regeneration are achieved. However, over time the morphogenetic competence of differentiated cultures declines (Lynch and Benson, 1991). Therefore, the primary goal of synthetic seed production is to produce somatic embryos that resemble more closely to the true seed embryo in storage and handling characteristics so that they can be utilized as a unit for clonal propagation and germplasm conservation. Synthetic seeds may be hydrated or dehydrated and may be quiescent or not. Encapsulation of micropropagules enables to satisfy the requirements of synthetic seed development. The gelling agents used for encapsulation for production of synthetic seeds act as protective cover. The encapsulated synthetic seeds also contain growth nutrients, plant growth promoting microorganisms (mycorrhizah, *Rhizobium*, etc.), and/or other biochemical constituents necessary for optimal embryo-to-plant development (Fig. 4.1). Fungicides may also be used in the liquid gelling medium to protect them during storage and germination *in vitro* and *in vivo*.

There could be a number of possible artificial seed systems, depending upon the type of artificial seeds produced, need of artificial seeds, the economic feasibility and it will vary greatly among species(Pond and Cameron, 2003). Different types of artificial seeds and their preparation methods are being elaborated in this chapter. Although Murashige had clearly envisioned the potential of this new tool in terms of clonal propagation, he also commented, "but to be applicable, this cloning method must be extremely rapid, capable of generating several million plants daily and competitive economically with the seed method". Differently to zygotic embryos, which are protected by a seed coat and have an access to the nutrients that are accumulated in the cotyledons or in the endosperm, somatic embryos are naked and dependent on the culture medium (differences of true seeds and artificial seed are being given in Table 4.1). Hence, it was soon evident that the artificial seeds had to be similar to the natural seed as much as possible, that is, it required the development of appropriate procedures for the encapsulation and the accumulation of storage compounds, creating an "artificial endosperm".

**Table 4.1.** Differences of true seeds and artificial seed

| Characters | True Seed | Artificial Seed |
|---|---|---|
| Definition | Botanically ***seed*** is a mature ovule along with its food storage, either ***endosperm*** or ***cotyledon***. The essential part is the embryo contained within the integuments, but it may be used less critically to describe planting materials. In terms of seed science, ***seed*** can be described as any propagating material either resulting from of sexual (true botanical seed) or asexual (cuttings, layering, bulb, tubers, rhizomes, micropropagated plantlets etc.) means is used for raising a crop. | ***Artificial seed*** could be defined as artificially encapsulated somatic embryos, shoot buds, cell aggregates or any other tissue that can be used for sowing as a seed and that possesses the ability to be converted into a plant under *in vitro* or *ex-vitro* conditions, and that retains this potential even after storage |
| Seed coat | Seed coat is prominent and well defined | No seed coat |
| Reserve food | Reserve food is in the form of endosperm (all the monocotyledonous plants) or cotyledon (dicotyledonous plants) | Food reserve is in the form of synthetic nutrient medium |
| Polarity | True seeds are bipolar (plumule/coleoptile and radicle) | Generally unipolar (somatic embryos, meristem, shoot tips etc.) |
| Production system | Produce naturally | Skill oriented and artificially produced |
| Uses | Used for plant propagation | Also used for plant propagation |

## 4.1. DESICCATED/DRY SYNTHETIC SEED

Synthetic seed was produced first time by Kitto and Janick (1982, 1985) involving carrot somatic embryos. They used polyoxyethylene, which is readily soluble in water, dries to form a thin film, does not support the growth of microorganisms and is non-toxic to the embryo, leading to the production of desiccated synthetic seed. Desiccation can be achieved either slowly over a period of one or two weeks sequentially using chambers of decreasing relatively humidity, or rapidly by unsealing the Petri dishes and leaving them on the bench overnight to dry. Thus, desiccated synthetic seed technology is not suitable to satisfy the objective of synthetic seed production.

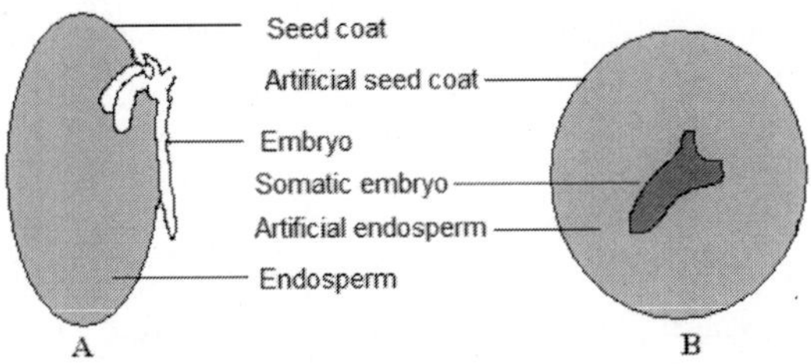


**Fig. 4.1.** Synthetic seed, gel encapsulated embryo with hydrophobic membrane. **A)** Cross-section of a dicotyledonous true seed, **B)** Cross-section of a synthetic seed.

### 4.1.1. Maturation of Embryos

Desiccated synthetic seeds are produced only in plant species whose somatic embryos are desiccation-tolerant. Before preparation of hydrated synthetic seeds, maturation of embryos is essential. 2,4-D has been successfully used in a large number of somatic embryogenesis studies. But, it is an inhibitor for precocious germination, whereby it allows embryo maturation to proceed in a more normal fashion, generally increasing the uniformity of the produced embryos and reducing the development of abnormal forms (Ammirato, 1983). ABA was also found to be critical for conversion of somatic embryos into plantlets (Redenbaugh *et al.*, 1991). If the embryos germinate, there will be insufficient time to accumulate storage reserves or acquire desiccation tolerance, and the embryo does not become quiescent. Imposing an osmotic stress on the embryos by supplementing medium with 5-6% sucrose instead of the normal 3%, prevents precocious germination and maintains embryogenic development. Accumulation of storage proteins and carbohydrates also induce desiccation tolerance (Lai and McKersie, 1993, 1994ab). If the nutrient requirement is satisfied by either addition of nutrients in the medium or by the frequent transfer to the fresh medium, somatic embryos will accumulate storage protein. The proportion of storage reserves in the embryo that accumulate starch or protein is regulated by the carbon: nitrogen ratio in the medium. Proteins accumulation may be enhanced by supplementing the medium with organic nitrogen sources, glutamine, and inorganic sulphur sources and potassium sulphate. Glutamine plays a regulatory as well as nutritive role in somatic embryo maturation. When included in the medium, glutamine converts to 5-oxoproline by autoclaving (Lai *et al.*, 1992; Lai and McKersie, 1994b).

### 4.1.2. Desiccation Tolerance

Desiccation tolerance can be achieved by addition of abscisic acid (ABA) in the medium during maturation stage of embryos and the cultures are kept on that medium for 3-4 days. Medium containing 20 mM ABA is sufficient to induce desiccation tolerance. This growth regulator apparently triggers a process leading to the expression of desiccation tolerance. Other physical treatments include cold, heat, osmotic or nutrient stresses, which have similar response. To induce desiccation tolerance through chilling, it requires three weeks.

McKersie *et al.* (1989) induced somatic embryos to acquire desiccation tolerance by treatment with ABA or any one of several environmental stresses, including water and nutrient stresses, applied to the embryoids at the cotyledonary stage of development. The embryoids were subsequently air dried slowly (over 7 days) or rapidly (over 1 day) to moisture contents of less than 15% and remained fully viable. Dry somatic embryos were stored with no loss of viability for 8 months at room temperature and humidity. On the contrary, Marsolair *et al.* (1991) described the desiccated somatic embryos as more suitable than somatic embryos encapsulated in hydrated gel for long term storage and for ease of handling, transport and sowing with automated equipment. Bornman *et al.* (2003) reported that the percentage germination of fresh or somatic embryos partially desiccated at relative humidity of 97% and 63% to moisture content approaching those of the seed, was substantially lower. This sensitivity to drying suggests that the somatic embryo seed may behave either as an orthodox seed with limited ability to withstand desiccation or as a recalcitrant seed that cannot survive drying below a moisture content that is relatively high.

Desiccation of somatic embryos provides a quiescent phase analogous to true seeds, facilitating the convenience of year round production, storage and distribution. Somatic embryos possess the ability to express desiccation tolerance in response to an external chemical or physical stimulus. The mechanisms of desiccation tolerance involves stabilization of membrane in dry state and prevention of oxidative degradation of bio-molecules.

### 4.1.3. Germination and Seedling Vigour

Dry somatic embryos lack the vigour normally which are associated with the seedlings from normal seeds. The reason for this is not yet been specified, although there are several possibilities. The dry somatic embryos may lack storage proteins or some other critical components required after germination by the seedlings. The storage protein levels have been increased with some improvement in vigour (Lai *et al.*, 1992; Lai and McKersie, 1993, 1994a & b). Somatic embryos store starch and sucrose, whereas, seeds store a hemicellulose in the cell wall of the endosperm called galatomannan (McCleary and Matheson, 1974, 1976). The sucrose reserves in the dry somatic embryos are rapidly depleted after imbibitions (Lai *et al.*, 1995). In some instances, there may be injury to the somatic embryos if the proper dying procedure is not followed (Lecouteux *et al.*, 1993).

In comparison to true seeds, water uptake during imbibitions of dry somatic embryos is quite rapid, because the somatic embryos lack a testa, there is no barrier to water uptake. Imbibition injury is, therefore, another possible cause of poor seedling vigour in synthetic seeds. Pre-hydration treatment of the somatic embryos in a moist atmosphere (100% relative humidity in a sealed chamber over water) for 24 hours improves the vigour. Although the storage protein and starch reserves are rapidly hydrolyzed following imbibition, the germination of dry somatic embryos on a nutrient medium greatly improve seedling development (Lai and McKersie, 1995). The most probable cause of poor seedling vigour is abnormal apical or plumule development. The dry synthetic seeds also showed much lower conversion rates.

### 4.1.4. Desiccated Artificial Seed Preparation Method

Fig. 4.2 showed schematic representation of desiccated synthetic seed production. Janick *et al.* (1993) have reported that coating a mixture of carrot somatic embryos and callus in polyoxyethelene glycol produced desiccated artificial seeds. The coating mixture was allowed to dry for several hours on a Teflon surface in a sterile hood. The dried mixture was then placed on a culture medium, allowed to rehydrate; and then scored for embryo survival.

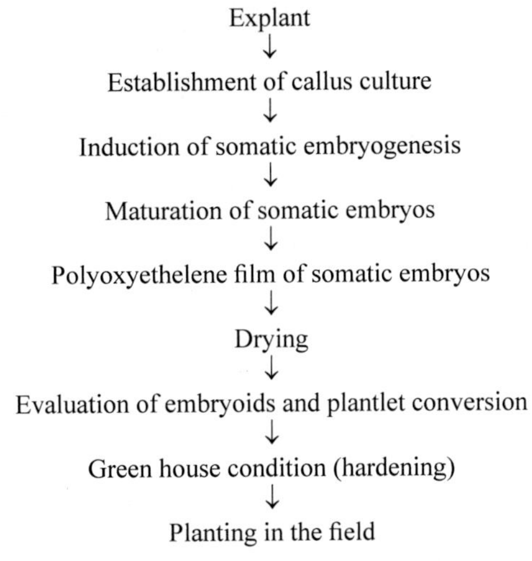


**Fig. 4.2.** Schematic representation of desiccated synthetic seed production from somatic embryos

## 4.2. HYDRATED SYNTHETIC SEED

Hydrated seeds are produced in those plant species where the somatic embryos are recalcitrant and sensitive to desiccation. Encapsulation of somatic embryos in hydrogel capsules produces hydrated synthetic seeds. The flow diagram of synthetic seed production using sodium alginate has been detailed in Fig. 4.3.

Propagules (somatic embryos, apical buds, axilary buds, immature seed embryo, protocorm-like bodies etc.)

↓

Mixing of propagules in sterile alginate with nutrients solution

↓

Individual propagules dropped into calcium salt solution

↓

Hardening of sodium alginate capsules

↓

Washing of capsules in sterile distilled water

↓

Evaluation for plantlet conversion

↓

Planting *in vivo* or *in vitro*

**Fig. 4.3.** Schematic representation of hydrated synthetic seed production using sodium alginate.

The most common used method to induce artificial seed is isnotropic gelation of sodium alginate by calcium ions. Redenbaugh *et al.* (1984) developed the technique of hydragel encapsulation of individual somatic embryos of alfalfa. Since then encapsulation in hydragel remains to be the most studied method of artificial seed production. A number of substances like, potassium alginate, sodium alginate, carregeenan, agar, gelrite, sodium pectate, etc. have been tested as hydragels (Table 3.1). Sodium alginate has a low cost and good gelation and biocompatibility characteristics. Smidsrod and Skjak-Braek (1990) suggested alginate for preparation of artificial seeds. Though many coating materials have been tried for encapsulation of somatic embryos, sodium alginate obtained from brown algae has been considered as the best and is being popularly used at present. Alginate has been chosen for case of capsule formation as well as for its low toxicity to the embryo. Alginate capsulated embryos can resist unfavourable field conditions without desiccation. The rigidity of the gel beads protects the fragile embryo during handling. These seeds so developed behave like true seeds and are used as substitute of natural seeds, which can be sown directly in the green house or in field.

### 4.2.1. Synchronization of Embryogenesis

A synchronized maturation of somatic embryo is very important to achieve good conversion ratio. Number of methods is available for selection of uniformity in developmental stage of calli. For suspension culture of carrot, all cells and cell aggregates of suspension culture are passed through sieve of particular mesh size (usually 100 mesh). Sieved cells or cell aggregates are incubated on gyratory shaker at 25 °C at 120 rpm. After 4 days, pro-embryos are available in the culture, and these pro-embryos are passed through 200 mesh sieve. Sieved pro-embryos will mature uniformly and synchronized embryogenesis may be obtained.

Two standard methods have been used so far for encapsulation of somatic embryos and other propagules.

### 4.2.2. Encapsulation by Dropping of Hydragel into Complexing Agent

Synthetic seed is produced by enclosing viable plant materials such as somatic embryos, androgenic embryos, pro-embryos, embryos-like-structure, protocorms, buds, etc. in alginate with nutrient sources. For encapsulation, the somatic embryos isolated are submerged in the sodium alginate solution (0.5-5%, w/v), according to the type of encapsulation applied, and subsequently suctioned through a micropipette to provide a protective capsule,which is prepared in suitable tissue culture basal medium supplemented with sucrose.

Embryos or buds are then suctioned through a micropipette individually and dropped into sterilized aqueous solution of 2-3% (w/v) calcium salt solution [$CaCl_2$ or $(CaNO_3)_2$] with occasional agitation. Calcium alginate beads were formed within 15-30 minutes. Here, ion-exchange reaction occurs and sodium ions are replaced by calcium ions forming calcium alginate beads or capsules surrounding the embryo/bud. This process is carried out under aseptic conditions in a laminar flow chamber, laminar with prior sterilization of the material and culture medium.The size of the bead depends upon the inner diameter of the pipette nozzle. Compactness and hardening of the encapsulated bead is modulated with concentration of sodium alginate and calcium chloride as well as duration of complexing. Optimum concentration of sodium alginate for production of synthetic seed ranges from 2-3% with a complexing solution containing 75-100 mM calcium chloride (Ara *et al.*, 1999; Bhattacharjee *et al.*, 1998; Bekheet *et al.*, 2002; Sparg *et al.*, 2002; Priya *et al.*, 2003). Gui *et al.* (1989) and Arun Kumar *et al.* (2005) used high concentration (4%) sodium alginate for artificial seeds. However, Molle *et al.* (1993) found that for the production of synthetic seeds of carrot, 1% sodium alginate, 50 mM $CaCl_2$ and 20-30 minutes soaking time period were satisfactory for proper hardening of calcium alginate capsules. They have suggested the use of a dual nozzle pipette in which the embryos flow through the inner pipette and the alginate solution through the outer pipette. As a result, the embryos are positioned in the centre of the bids for better protection. Preparation of hydrated synthetic seed using sodium alginate has been schematically presented in Fig. 4.4. So, to prevent above problems, one or two more coatings of alginate are given and also, certain antibiotics, insecticides, pesticides, etc. are added to prevent the microbial attack.

### *4.2.2.1. Preparation of sodium alginate*

- Take 500 ml capacity conical flask.
- Weigh 6 g sodium alginate.
- Take 200 ml of ½ MS basal in the conical flask.
- Heat the medium on a magnetic stirrer with hot-plate.
- Add sodium alginate in the heating MS medium and dissolve the sodium alginate with nutrient medium using magnetic stirrer. It may take 30-40 minutes for complete melting of sodium alginate.
- Close the mouth of the conical flask and autoclave it for 15 minutes. Thus, sodium alginate is ready for use.

### *4.2.2.2. Preparation of calcium chloride*

- Take 500 ml capacity conical flask.
- Weigh 6 g calcium chloride and pour into the conical flask.
- Dissolve it in distilled water.
- Makeup the volume to 200 ml.
- Close the mouth of the conical flask and autoclave it for 15 minutes. Thus, sodium alginate is ready for use.

### *4.2.2.3. Practical steps for synthetic seed preparation*

- Mix the sterilized encapsulating material (plant propagules).
- Take calcium chloride into autoclaved beaker of 100 ml capacity.
- Take one plant propagule along with sodium alginate with the help of dropper.
- Drop it into the calcium chloride.
- Continuously shake the beaker holding calcium chloride.
- Keep the beads in calcium chloride for 15-20 minutes.
- Decant the calcium chloride.
- Wash the beads with autoclaved distilled water to remove calcium chloride from the surface of synthetic seeds. Thus, the synthetic seeds are ready for preservation or transport or germination.

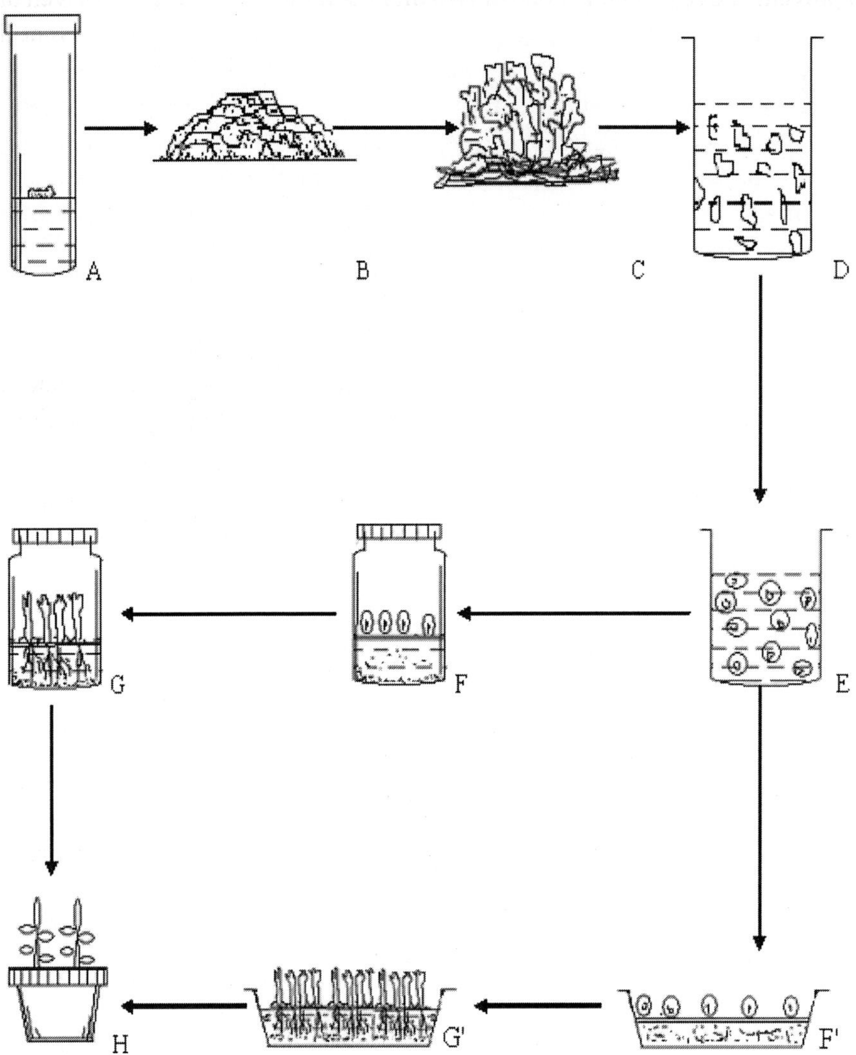


**Fig. 4.4.**Flow diagram presenting the stepwise synthetic seed production from somatic embryos.**A)** Explant on culture medium, **B)** Callus growth, **C)** Maturation followed by somatic embryo formation, **D)** Isolated embryos mixed with sodium alginate, **E)** Encapsulation in calcium chloride solution, **F)** Testing of embryos for plantlet regeneration, **G)** Germinated plants, **F′)** Beads on vermiculite in greenhouse for germination, **G′)** Germination of beads on vermiculite, **H)** Grown up plant in pot in *in vivo* condition.

## 4.2.3. Partially Encapsulated Artificial Seed

This technique of preparation of synthetic seed was developed by Mamiya and Sakamoto (2001). Empty Ca-alginate beads (i.e., without explants) are prepared as that of conventional "drop-by-drop" release of alginate into the complexing calcium agent.a hole is made in the capsules from which the explant (a somatic embryo) is then inserted to form the synseed. An advantage of this method is related to the fact that the embryo primordium is not embedded totally in the alginate matrix, thus preventing any possible physical inhibition to the embryo conversion to plantlet. However, the complexity of the technique, in comparison with the conventional "drop-by-drop" method, is a weak point which has prevented its diffusion.

### 4.2.4. Ca-alginate Hollow Beads

This technique was developed by Patel *et al.* (2000). Here, the explants (i.e., cells or shoot tips of potato and carrot) were suspended in a solution containing carboxymethylcellulose and calcium chloride, and then dripped into a Na-alginate solution. The technique allowed the explants to regrow (cells) or to be converted to shoots (shoot tips) directly inside the capsules. The advantage of this new concept is that, in contrast to the conventional method where the explant is often located near the surface, here the explant is positioned exactly at the centre of the bead, thus ensuring the complete protection of the embryo (or other explant) as in the natural seed.

### 4.2.5. Molding

In this, embryos are mixed with temperature dependent gel like gelrite where cells get gelled as the temperature is lowered. However, the synthetic seeds obtained by above methods cannot be used directly into the fields because:

- They are very fragile
- Their nutrients will be attacked by the microflora present in the soil which will starve the embryo to death

### 4.2.6. Self-breaking Artificial Seed

In encapsulation technology, there may be physical hindrance of shoot or root emergence caused by the gel capsule. This can be overcome by adopting self-breaking alginate gel bead technology in which artificial seeds are pretreated with potassium nitrate ($KNO_3$) in which, during the pretreatment with $KNO_3$, the $K^+$ ions replace the $Ca^{2+}$ ions of the calcium alginate capsule thus allowing the artificial seeds to soften and allow the subsequent conversion to plantlets (Onishi *et al.*, 1994). This self-breaking artificial seeds technology was successfully applied to *Feijoa sellowiana* (goiabeiraserrana), *Oryza sativa* (hybrid rice) and *Stevia rebaudiana* (stevia) (Guerra *et al.*, 2001; Arun Kumar *et al.*, 2005; Ali *et al.*, 2012).

Simply encapsulated embryos in many cases do not yield the desired conversion rate in the field probably due to the elastic nature or strength of the gel beads and the oxygen deficiency inside gel beads. The development of synthetic seed involving the production of artificial endosperm and self-breaking treatment for encapsulation and sorting were described by Sakamoto *et al.* (1992). Self-breaking synthetic seeds are prepared following the steps as given below:

- The gel beads are rinsed thoroughly with soft water/tap water for 40 minutes
- Immersed them in a monovalent cation solution of potassium nitrate of 200 mM for 60 minutes.
- Another 40 minutes washing is required for desalting.
- The gel beads then gradually swell and split within six hours if sown in humid condition.

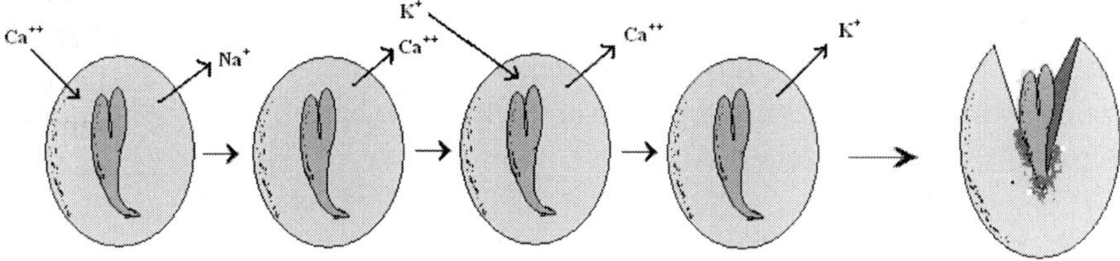


**Fig. 4.5.** Different phases of self-splitting synthetic seed

### 4.2.7. Double Layered Artificial Seed

Recently technology has been developed to prepare double-layered synthetic seeds (Kinoshita, 2002). For somatic embryo encapsulation, sodium alginate is largely used; however, this is excessively permeable with loss of the nutritive substances and/or dehydration risk during conservation and transport causing detrimental effects on the synthetic seed conversion and the plantlet growth. In order to overcome these problems, Micheli *et al.* (2002) developed double coat encapsulation and encapsulation coating procedure in M.26 apple rootstock. The inner layer contained a large quantity of sucrose. To prevent the diffusion of sucrose from artificial seeds to non-sterilized substrate, artificial seeds were enveloped in a dialysis membrane. The enveloped artificial seeds germinated quickly in non-sterilized vermiculite.

An additional coating on sodium alginate capsule, like the Elvax polymer coating has been used (Redenbaugh *et al.*, 1988). Following are the main advantage of double layered synthetic seed:

- Outer layer protect the inner soft-nutrient layer from mechanical injury
- Fungicide can be added to the second layer to prevent fungal infection
- To stop diffusion of sucrose from the gelling matrix of first layer
- It facilitates *in vivo* sowing of synthetic seeds
- It helps long transportation
- Improves storability leading in long term conservation of plant materials

#### *4.2.7.1. Coating of mineral oil*

Mukunthakumar and Mathur (1992) reported an increase of 56% of the *in vivo* plantlets conversion following an additional coating of mineral oil on the alginate beads of *Dendrocalamusstrictus*. This mineral oil layer protects against desiccation, tackiness and invading microbs. The mineral oil may even be used to store synthetic seeds for six months or more (Mathur *et al.*, 1991).

### 4.2.8. Automated Encapsulation

The first report of large-scale embryogenic culture described an attempt to grow carrot callus in 20-litre carboys, which resulted in the formation of few embryos (Backs-Husemann and Reinert, 1970). The biological/mechanical system most often described for application to embryogenic system is the stirred-tank bioreactor, a mass culture system originally developed for microbial fermentation, but now adopted for growing plant cells on a large scale (Martin, 1980; Kurz and Constabel, 1981). This large-scale production of somatic embryos can be used for preparation of synthetic seeds.

Automated encapsulation process is a quick method of artificial seed production. An encapsulation machine can be used successfully to encapsulate somatic embryo in order to achieve a time and hand labour saving and to increase the accuracy. Sicurani *et al.* (2001) developed mechanical excision of explants. This technique is useful for the production of synthetic seeds through encapsulation of differentiating propagules (tissue fragments with shoot primodia) in woody species. Brischia *et al.* (2002) also used mechanically manipulated explants of apple rootstock for encapsulation. They suggested that machine processed explants can be encapsulated for production of synthetic seeds.

The development and preservation of embryogenic cell lines, somatic embryogenesis in bioreactors and embryo-to-plant conversion; preservation and coating were discussed by Petiard *et al.* (1993) with reference to carrots and *Coffea camphora*. About 55000 and 5000 embryos/litre can be produced daily for carrots and coffee, respectively. Here the economics of mass somatic embryogenesis (artificial seed production) of coffee trees were considered. Alternatively, the embryos could be mixed in a temperature depended gel such as, gelrite, agar, agarose etc., placed in the well of a micro-titer plate, and gelled as the teperature was lowered. The judicial and intelligence coupling of artificial seed technology with that of microcomputer in achieving automated encapsulation and regeneration of plantlets would tremendously increase the efficiency of encapsulation and production of homogeneous and high quality seeds, and will thus revolutionize the current concept of commercial micropropagation method.

### 4.2.9. Semi-automated Encapsulation

Somatic embryos are mixed in sodium alginate solution (2-5%) prepared with the appropriate nutrient medium. Sodium alginate solution impregnated with somatic embryos is dropped into calcium chloride solution (Fig. 4.6). Other steps remain the same as described in section 4.2.2.

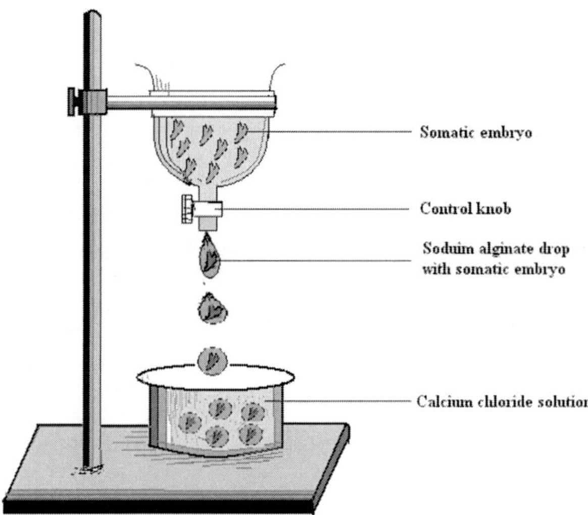


**Fig. 4.6.**Diagrammatic representation of semi-automated encapsulation of somatic embryos

Molle *et al.* (1993) have suggested the use of a dual nozzle pipette in which the embryos flow through the inner pipette and the alginate solution through the outer pipette. As a result, the embryos are positioned in the centre of the beads for better protection.

## 4.3. EFFECT OF NUTRIENTS ON PERFORMANCE OF ARTIFICIAL SEED

It has been shown that the nutrients to be included in the alginate matrix should be adapted according to the type of propagule and plant species (Naik and Chand, 2006). However, the complex composition of the endosperm makes difficult its reconstitution and the use of standard culture medium in the capsule may increase the conversion of the propagule to plantlets. It was also noted that the addition of extra $KNO_3$ to MS medium increased microshoots conversion (Sandoval-Yugar*et al*., 2009). Their results showed that the capsule weakening through $KNO_3$ treatment resulted in high conversion capacity of the shoots, which is improved by an additional nutrient supply.

Contamination of synthetic seeds was low in the absence of sucrose and showed higher *in vitro* conversion capacity as compared to sucrose-enriched gelling matrix (Sandoval-Yugar *et al*., 2009). Probably in the absence of sucrose, the microshoots were forced to undergo photosynthesis. It was well established that a regulative feedback occurs in response to the presence of sugar and the photosynthetic capacity of the plant. In contrast, Matsumoto *et al*. (1995) reported higher *in vitro* conversion in presence of sucrose in the culture medium.

## References

Ali A, Gull I, Majid A, Saleem A, Naz S, Naveed NH. 2012. *In vitro* conservation and production of vigorous and desiccate tolerant synthetic seeds in *Stevia rebaudiana*. J Med Plants Res. 6: 1327-1333.

Ammirato PV. 1983. The regulation of somatic embryo development in plant cell culture: Suspension culture technique and hormone requirements. BioTechnol. 1: 68-74.

Ara H, Jaiswal U, Jaiswal VS. 1999. Germination and plantlet regeneration from encapsulated somatic embryos of mango (*Mangiferaindica* L.).Plant Cell Rep.19: 166-170.

Arun Kumar MB, Vakeswaran V, Krishnasamy V. 2005. Enhancement of synthetic seed conversion to seedlings in hybrid rice. Plant Cell Tiss Org Cult. 81: 97-100.

Backs-Husemann D, Reinert J. 1970. EmbryobildungdurchisolierteEinzellenausGewebkutturen von *Ducuscarota*.Protoplasma. 70: 36-90.

Bekheet SA, Taha HS, Saker MM, Moursy HA. 2002. A synthetic seed system of date palm through somatic embryos encapsulation. Annals AgretSci Cairo. 47(1): 325-337.

Bhattacharjee S, Khan HA, Reddy PV, Bhattacharjee S. 1998.*In vitro* seed germination, production of synthetic seeds and regeneration of plantlets of *Phalaenopsis*hybrid. Annals AgrilSci, Cairo. 43(2): 539-543.

Bornman CH, Dickens OSP, Merwe CF-Van-der, Coetzee J, Botha AM, der-Merwe CF-van, Van-der-Merwe CF. 2003. Somatic embryo of *Piceaabier* behave like isolated zygotic embryos *in vitro* but with greatly reduced physiological vigour. South African J Bot. 69(2): 176-185.

Brischia R, Piccioni E, Standardi A. 2002. Micropropagation and synthetic seed in M.26 application root stock (II): A new protocol for production of encapsulated diffentiating propagules. Cell Tiss Org Cult. 68(2): 137-141.

Guerra MP, Dal Vesco LL, Ducroquet JPHJ, Nodari RO, Reis MS. 2001. Somatic embryogenesis in *Goiabeiraserrana*: genotype response, auxinic shock and synthetic seeds. Braz J Plant Physiol. 13: 117-128.

Gui YL, Xu TY, Gu SR, Guo ZS, Hou SS, Ke SG, Wu YL, Li HL. 1989. Studies on somatic embryogenesis of *Coptischinensis.* Acta Botanica Sinica 31(12): 923-927.

Janick J, Kim YH, Kitto S, Saranga Y. 1993. In: Synseeds, Redenbaugh, K (ed.), CRC Press, Boca Raton, pp. 12-34.

Kinoshita L. 2002. Artificial seeds enveloped with dialysis membrane germinated quickly under non-aseptic conditions. Bulletin of the Forestry and Forest Products Res Inst, Ibaraki. 385: 241-244.

Kitto SK, Janick J. 1982. Polyox as an artificial seed coat for asexual embryos.Horti Sci. 17: 488

Kitto SK, Janick J. 1985. Production of synthetic seeds by encapsulating asexual embryos of carrot. J AmerHort Sci. 110: 277-282.

Kurz WGW, Constabel F. 1981. Continuous culture of plant cells. In: Continuous Cultures of Cells, Vol. 2, Calcott PH (ed.). CRC Press, Boca Raton. pp. 141-157.

Lai FM, Lecouteux CG, McKersie BD. 1995. Germination of alfalfa (*Medicago sativa* L.). I. Subcultures and indirect secondary somatic embryogenesis. Plant Cell Tiss Org Cult. 37: 151-157.

Lai FM, McKersie BD. 1993. Effect of nutrition on maturation of alfalfa (*Medicago sativa* L.) somatic embryos. Plant Sci. 91: 85-87.

Lai FM, McKersie BD. 1994a. Regulation of starch and protein accumulation in alfalfa (*Medicago sativa* L.) somatic embryos. Plant Sci. 100: 211-219.

Lai FM, McKersie BD. 1994b. Regulation of storage protein synthesis by nitrogen and sulpher nutrients in alfalfa (*Medicago sativa* L.) somatic embryos. Plant Sci. 103: 209-221.

Lai FM, McKersie BD. 1995. Germination of alfalfa (*Medicago sativa* L.) seeds and somatic embryos. II. Effects of nutrient supplements. J Plant Physiol. 146: 731-735.

Lai FM, Senaratna T, McKersie BD. 1992. Glutamine enhances storage protein synthesis in *Medicago sativa* L. somatic embryos. Plant Sci. 87: 69-77.

Lecouteux C, Lai FM, McKersie BD. 1993. Maturation of somatic embryos by abscisic acid, sucrose and chilling. Plant Sci. 94: 207-213.

Lynch PT, Benson EE. 1991. Cryopreservation, a method for maintaining the plant regeneration capacity of rice, cell suspension culture. In: Proceedings of the Second Rice Genetic Symposium, International Rice Research Institute, Los Banos, Philippines. pp. 321-332.

Mamiya K, Sakamoto Y. 2001. A method to produce encapsulatable units for synthetic seeds in *Asparagus officinalis*. Plant Cell Tiss Org Cult.64: 27-32.

Marsolair AA, Wilson DPM, Tsujita MJ, Seneratna T. 1991.Somatic embryogenesis and artificial seed production in zonal (*Pelargonium hortorum*) and regal (*Pelargonium domesticum*) geranium. Canadian J Bot. 69(6): 1188-1193.

Martin SM. 1980. Mass culture system for plant cell suspesion. In: Plant Tissue Culture as a Source of Biochemicals, Stabba EJ (ed.). CRC Press, Boca Raton. pp. 149-166.

Mathur J, Mukunthakumar S, Gupta SN, Mathur SN. 1991. Growth and morphogenesis of plant tissue cultures under mineral oil. Plant Sci. 74: 249-254.

Matsumoto K, Matsumoto K, Hirao C, Teixeira J. 1995. *In vitro* growth of encapsulated shoot tips in banana (*Musa* sp.). ActaHorticulturae. 370: 13-20.

McCleary BV, Matheson NK. 1974. Galactomannan activity and galactomannan and galactosyl-sucrose oligosacchride depletion in germinating legume seeds. Phytochem. 13: 1747-1757.

McCleary BV, Matheson NK. 1976. Galactomannan utilization in germinating legume seeds. Phytochem. 15: 43-47.

McKersie BD, Senaratna T, Bowley SR, Brown DCW, Krochko JE, Bewley JD. 1989. Application of artificial seed technology in the production of hybrid alfalfa (*Medicago sativa* L.). In Vitro Cell Dev Biol. 25: 1183-1188.

Micheli M, Pellegrino S, Piccioni E, Standardi A. 2002. Effect of double layered encapsulation and coating on synthetic seed conversion in M.26 apple rootstock.J Microencapsulation. 19(3): 347-356.

Molle F, Dupius JM, Ducos JP, Anselm A, Cerolus SI, Petiard V, Freyssinet G. 1993. In: Synseeds, Redenbaugh K (ed.). CRC Press, Boca Raton. pp. 257-270.

Mukunthakumar S, Mathur J. 1992. Artificial seed production in the male bamboo *Dendrocalamusstrictus* L. Plant Sci. 87: 109-113.

Naik SK, Chand PK. 2006. Nutrient-alginate encapsulation of *in vitro* nodal segments of pomegranate (*Punicagranatum* L.) for germplasm distribution and exchange. ScientiaHorticulturae. 108: 247-252.

Onishi N, Sakamoto Y, Hirosawa T. 1994. Synthetic seeds as an application of mass production of somatic embryos. Plant Cell Tissue Org Cult. 39(2): 137-145.

Patel AV, Pusch I, Mix-Wagner G, Vorlop KD.2000. A novel encapsulation technique for the production of artificial seeds. Plant Cell Reports.19: 868-874.

Petiard V, Ducos JP, Florin B, Lecouteux C, Tessereau H, Zamarripa A. 1993. Mass somatic embryogenesis: a possible tool for large-scale propagation of selected plants. In: Proceedings of the Fourth International Workshop on Seeds: Basic and Applied Aspects of Seed Biology. Agers, France, 20-24 July, 1992. pp. 175-191.

Pond S, Cameron S. 2003.Artificial Seeds, Tissue Culture, Elsevier Ltd. pp. 1379-1388.

Priya BT, Arumugam Shakila, Shakila A. 2003. Synthetic seed production in banana: Adv Plant Sci. 16(1): 219-222.

Redenbaugh K, Fujii JA, Slade D. 1988. Encapsulated plant embryos. In: Biotechnology in Agriculture, Mizrahi A (ed.). Alan R. Liss, Inc., New York. pp. 225-248.

Redenbaugh K, Fujii JA, Slade D. 1991. Synthetic seed technology. In: Cell Culture and Somatic Cell Genetics in Plants, Vol. 8, Vasil, I.K. (ed.), Academic Press, New York. pp. 35-74.

Redenbaugh K, Nichol J, Kosseler ME, Paasch BD. 1984. Encapsulation of somatic embryos for artificial seed production. *In Vitro* Cell DevBiol Plant. 20: 256-257.

Sakamoto Y, Mashiko T, Suzuki A, Kawata H, Iwasaki A, Hayashi M, Kano A, Goto E. 1992. Development of encapsulation technology for synthetic seeds. ActaHorticulturae. 319: 71-76.

Sandoval-Yugar EW, Vesco LLD, Steinmacher DA, Stolf EC, Guerra MP. 2009. Microshoots emasculation and plant conversion of *Musa* sp. cv. 'Grand Naine'. Ciência Rural Santa Maria. 39(4): 998-1004.

Sicurani M, Piccioni E, Standardi A. 2001. Micro-propagation preparation of synthetic seed in M.26 apple root stock I: attempts towards saving labour in the production of adventitious shoot tips suitable for encapsulation. Plant Cell Tiss Org Cult. 66(3): 207-216.

Smidsrod O, Skjak-Braek G. 1990. Alginate as immobilization matrix for cells.Trend Biotechnol. 8(3): 71-78.

Sparg SG, Jones NB, Staden J-van, Van-Staden J. 2002. Artificial seed from *Pinuspatula* somatic embryos. South African J Bot. 68(2): 234-238.

Vdovitchenko YM, Kuzovkina IN. 2011. Artificial Seeds as a Way to Produce Ecologically Clean Herbal Remedies and to Preserve Endangered Plant Species. Moscow Univ Biolog Sci Bullet66(2): 48-50.

# 5

# Propagules for Encapsulation

As stated by Murashige's definition, artificial seed (or synthetic/synseed) technology was initially restricted to species in which somatic embryogenesis was possible. Later, Bapat and co-workers (1987) proposed to broaden the technology to the encapsulation of various *in vitro*-derived propagules, and they used axillary buds of *Morus indica* as a first example of this new application. The new concept paved the way for the encapsulation of explants other than somatic embryos, and to the formulation of a new definition of artificial seeds (Aitken-Christie *et al.* 1995) as "artificially encapsulated somatic embryos, shoots, or other tissues which can be used for sowing under *in vitro* or *ex vitro* conditions". In other words, the "new" artificial seed, lost the original bond of containing an embryo (zygotic or somatic), can now contain any kind of explant from tissue culture (such as axillary buds, shoot tips, nodal segments, bulblets, protocorms, callus samples, cells) which, following 'germination', will evolve into a plantlet or a shoot (artificial seed conversion) or will produce new cell proliferation (artificial seed regrowth).

Realization of synthetic seed technology requires manipulation of *in vitro* culture system for large-scale production of viable materials for preparation of synthetic seed, subsequently conversion into complete plants under *in vitro* as well as *in vivo* conditions. Somatic and gametic embryogenesis that enhanced axillary bud proliferation are the efficient techniques for rapid and large-scale multiplication of desirable plant species. Plant tissues such as somatic embryos, apical shoot tips, axillary shoot buds, embryogenic calli, and protocom-like bodies are potential micropropagules that have been considered for creating synthetic seeds. Encapsulations of somatic embryos, apical and axillary shoot buds, and regeneration of whole plants from them have been reported for a number of plant species (Redenbaug *et al.*, 1986; Mathur *et al.*, 1989; Ganapathi *et al.*, 1992; Lulsdorf *et al.*, 1993).

## 5.1. EMBRYOGENIC SYNTHETIC SEEDS

The process of somatic embryogenesis, in which the somatic cells or tissues develop into differentiated embryos, produces somatic embryos and each fully developed embryo is capable of developing into a plantlet. Embryos can be obtained either directly from cultured explants, anther or pollen, callus and isolated single cell in culture. The concept of somatic embryogeny and the production of high quality embryos is not one that is widely followed by many researchers, who have either focused on the production of a somatic embryogenesis system, the result in some plants recovery or who have interested only in the study of the early stage of embryogenesis. This focused on production of matured and true-to-type high quality embryos which can possibly lead to the conclusions based on abnormal somatic embryogenesis. To achieve high quality embryo production for scale-up of artificial seed production technology, measures has to be taken to improve the embryo response in terms of embryo-to-plant development.

Somatic embryos are bipolar, under optimum culture medium it produces root and shoot simultaneously and this process is known as ***somatic embryogenesis***. For synthetic seed production, embryos may be obtained from the following sources.

### 5.1.1. Concept of Totipotency

***Totipotency*** is the ability of a cell, tissue or organ to grow and develop into a multi-cellular or multi-organed higher organism. In natural state, plant cell does not have the totipotency. Few exceptions, leaves of bryophyllum can regenerate into whole plant. Some tissues do not divide at all, other do so only occasionally. Cell or tissue or organ under appropriate conditions can differentiate into a whole plant. The meristem tissues divide and grow into complete plant(s) under appropriate environment and medium. The conversion of meristematic tissue into specialized tissues with specific function is termed as ***differentiation***. They are used for clonal propagation whereby new plants are generated through organogenesis.

Plant cell or tissue has the ability to multiply into mass of undifferentiated cells, which is termed as *dedifferentiation*, and subsequently can undergo morphogenesis to regenerate a complete plant with all necessary organs of the mother plant is termed as *redifferentiation*. These three morphogenetic phenomena are being explained using the flow chart as given in Fig. 5.1. Optimum micro-environment is to be created under *in vitro* culture condition for the morphogenesis of the cell. These cell derived plants are morphologically and functionally similar among themselves as well as to the mother plant if somaclonal variation is ignored.

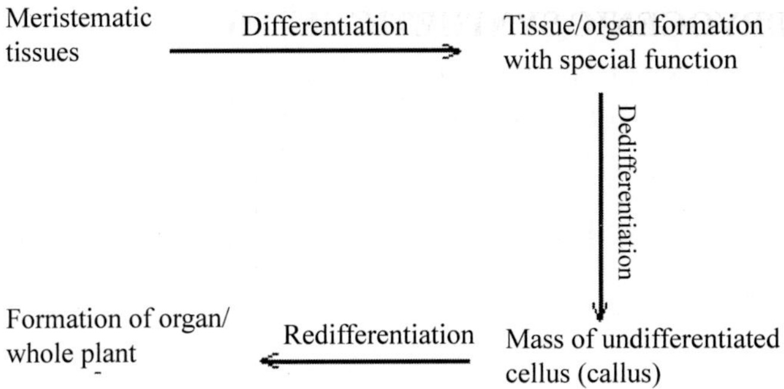


**Fig. 5.1.** Morphogenetic phenomena of *in vitro* culture of plants

## 5.1.2. Somatic Embryos

One of the prerequisite for the application of synthetic seed technology in *micropropagation*, is the production of large numbers of high-quality, vigorous somatic embryos that can produce plants with frequencies comparable to natural seeds. On tissue culture, the plant explant undergoes morphogenetic changes though two paths, namely *organogenesis* and *embryogenesis* (Fig. 5.2). Here we have more interest on somatic embryogenesis. Somatic embryogenesis is the process by which the somatic cells or tissues develop into differentiated embryos, and each fully developed embryo is capable of developing into plantlet. Somatic embryos are bipolar structures with apical and basal meristematic regions developed through tissue culture, which is capable to grow into a whole plant with root and shoot. Embryo can be obtained either directly from cultured explants or indirectly through callus culture.

Organogenesis leads to many other terminologies. Initiation of only adventitious shoot bud from callus is known as *caulogenesis*. Similarly, *rhizogenesis* is a type of organogenesis by which only root initiation takes place in the

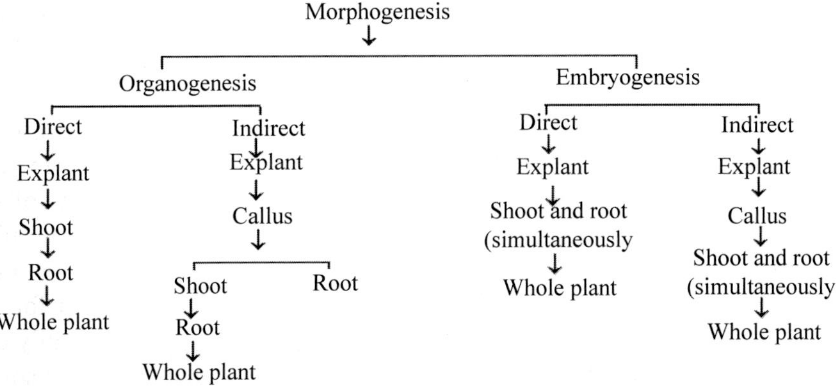


**Fig. 5.2.** Different paths of *in vitro* morphogenesis of plant

callus mass. Irregular structures when develop during organogenesis is called *organoids*. The localized meristematic cells on a callus which gives rise to shoots and/ or roots are termed as *meristemoids*.

Synthetic seed technology requires the less expensive production of large numbers of high-quality somatic embryos. Most of the researcher's interest in embryo formation using tissue culture techniques simply rest on explanation of regeneration of plantlets through somatic embryos or stages of embryogenesis. This focus separated from an emphasis on producing mature, true-to-type, high quality embryos that can possibly lead to conclusions based on abnormal somatic embryogenesis. Simply emphasizing the number of embryos per gram of fresh callus or the number of embryo-producing calli is not adequate. In fact, treatment that lead to production of higher numbers of embryos may actually produce fewer superior quality embryos that another protocol, which produces comparatively lower numbers of embryos but high percentage of quality embryos. To overcome this problem and to achieve embryo production for scale-up of artificial seeds, measure the embryo response in terms of embryo-to-plant development. It is important to produce large quantities of high quality somatic embryos comparatively in short period of time is very essential. Good quality embryos will increase the conversion rate of somatic embryos into plantlets subsequently the establishment of plant in *in vitro* or *in vivo*.

The quality of the artificial seed depends on the temporal qualitative supply of growth regulators and nutrients along with an optional physical environment. The advantages of preparing synthetic seeds from somatic embryos have been discussed by many authors (Gray and Purohit, 1991; Senaratna, 1992; Flacinelli *et al.*, 1993; Ara *et al.*, 2000). Plant propagation via somatic embryogenesis has been proposed for synthetic seed production of many crops (see Table 8.1). The use of somatic embryos as artificial seeds is becoming more feasible as the advances in tissue culture technology define the conditions for induction and development of somatic embryos in an increasing number of plant species (Jain *et al.*, 1995; Ipekci and Gozukirmizi, 2003). Various types of artificial seeds have been prepared using somatic embryos, which have been either dried (Senaratna *et al.,* 1989; Kim and Janick, 1990; McKersie *et al.,* 1989, Takahata *et al.*, 1992) or maintained fully hydrated (Ghosh and Sen, 1994; Padmaja *et al.*, 1995; Onay *et al.*, 1996; Xing *et al.*, 1995; Datta *et al.*, 1999, Priya *et al.*, 2003). The basic steps involve in production of somatic embryos for preparation of synthetic seeds are being brieflydiscussed.

### *5.1.2.1. Establishment of Aseptic Culture*

Explants (seeds, shoot tips, buds, leave, flower buds, immature inflorescence etc.) are cleaned and immersed in 5% aqueous Teepol solution (or any liquid detergent) with 1-2 drops of Tween20 as wetting agent under continuous shaking @ 150 rpm. These were thoroughly washed in distilled water (4 times) to remove excess detergent and surface sterilized with freshly prepared 0.1% $HgCl_2$ (concentration may vary form 0.1-1.0% depending on extent of contamination and the hardiness of the explant) for 15 minutes *in vacuo* followed by three wash in sterile distilled water to remove traces of $HgCl_2$ completely. Explants are blot dried under laminar air flow cabinet on sterile tissue paper and cultured on medium mentioned specifically under individual experiment. Many other surface sterilizing agents are available (Table 2.2). Among these, $HgCl_2$ is most frequently used.

### *5.1.2.2. Induction of Somatic Embryo*

After the discovery of somatic embryogenesis in 1950, it was possible to have an alternative of conventional zygotic seeds. Somatic embryo arises from the somatic cells of a single parent. They differ from zygotic embryos since somatic embryos are produced through *in vitro* culture, without nutritive and protective seed coats and do not typically become quiescent. Somatic embryos are structurally equivalent to zygotic embryos, but are true clones, since they arise from the somatic cells of a single parent.

Somatic embryogenesis offers as an ideal system for the production of somatic embryos on a large scale for use in the preparation of synthetic seeds. Somatic embryos can be induced through standard tissue culture techniques. Callus can be induced to dedifferentiate on an appropriate medium to develop embryoids and gradually mature into globular shaped embryos. Most important factors affecting the induction of embryogenic callus and plant regeneration through somatic embryogenesis include the explant type, media composition, growth regulators and other organic as well as inorganic adjuvant. Among these, growth regulator is the most important.

Induction of somatic embryogenesis requires a change in the fate of a somatic cell. In most cases, an inductive treatment is required to initiate cell division and establish a new polarity in the somatic cell. The most common callus induction treatments is 2,4-D (2,4-dichlorophenoxy acetic acid), but other auxins such as 2,4,5-T (2,4,5-trichlorophenoxy acetic acid), *p*-chlorophenoxy acetic acid, piclram (4-amino-3,5,6-trichloropicolinic acid) are effective. Inorganic components in the medium such as potassium and organic components like proline can modulate the embryogenesis or callus response, but they can not replace auxin.

Somatic embryogenesis is a genetic trait that is sexually heritable. Explants, depending upon the plant species or based on requirement of the type of callus are surface sterilized and cultured on any standard medium supplemented with 2,4-D or any other callus inducing growth regulators. Callus inducing growth regulators, namely 2,4-D activate the cell cycle of many cells of explants those in vascular cambium develop into callus, whereas some epidermal cells develop into a somatic embryo. The initial somatic embryos are only small dense cell clusters. These non-differentiated cell mass are embedded to prepare synthetic seeds. To liberate these proembryonic structures, and to stimulate the formation of more embryos, the callus may be dispersed in a standard liquid medium containing 2,4-D but not kinetin. Further, the medium can be modified as per requirement or based on the experience of the researchers. High levels of sucrose, nitrogen and sulphur could be added to the medium to prevent precocious germination of embryos (Anandarajah and McKersie, 1990b) and to enable storage reserves (Lai and McKersie, 1994a & b). To induce desiccation tolerance, somatic embryos may be placed on medium containing abscisic acid for 3-4 days (Senarantna *et al*., 1990).

Flow diagram presenting the procedure of synthetic production using somatic embryos was given in Fig. 4.4. Kamada *et al*. (1989) induced somatic embryos on the surface of the apical meristem of carrot cv. US-Harumakigosum seedling without visible callus formation were cultured on hormone free MS (Murashige and Skoog, 1962) medium with 0.7M sucrose or 0.25-1.00 mM Cd ion, then transferred to hormone-free MS medium with 0.1M sucrose. Passing fractionated these embryos through stainless steel sieves with different pore sizes were encapsulated in calcium alginate gel. These synthetic seeds germinated in 1-2 weeks after culturing. Chee and Contiliffe (1992) developed a protocol to produce heterogeneous population of embryogneic aggregates from sweet potato cv. White star for synthetic seed production. Embryogenic callus disassociates in liquid medium to form a heterogeneous population of embryogenic and non-embryogenic cell aggregates of varying sizes. To improve embryo production, such cell aggregate populations were obtained by manual fragmentation of calli. Embryogenic callus and embryo production per mg of cultured embryogenic callus increased quadratically with decreasing aggregate size.

### 5.1.2.3. Maturation of Somatic Embryos

The somatic embryo needs to reach a stage of necessary vigour during the maturation phase, which can break the mechanical resistance of the gelling matrix. Induction by an osmotic stress on the embryos by supplementing medium with 5-6% sucrose instead of the normal 3% prevents precocious germination and maintains embryogenic development.

### 5.1.2.4. Stage of Somatic Embryos for Preparation of Synthetic Seed

The process of embryogenesis involves various stages of differentiation and development such as ***proembryo***, globular, heart-shape and torpedo embryo. Friable cream colouredcalli are selected for induction of somatic embryos.

In case of monocot-plants, globular staged somatic embryos are suitable for encapsulation. Light brown friable calli are non-embryogenic, green compact with low embryogenic frequency, whereas yellowish-nodular friable calliposses high embryogeic frequency.

#### *5.1.2.5. Somatic Embryos versus Zygotic Embryos*

Both the somatic and zygotic embryos are structurally almost similar and have the ability to develop into a complete plant. The contrasting characters of these two types of embryos are being tabulated in Table 4.1.

**Table 4.1.** Distinguishing features of somatic and zygotic embryos

| Feature | Zygotic embryo | Somatic embryo |
|---|---|---|
| Definition | A young sporophytic plant, as yet retained in the gametophyte in the seed | Formation of embryo under *in vitro* condition from asexual cells |
| Initiation | Natural seed develops as a result of a sexual process i.e. pollination followed by fertilization | Somatic embryo develops from unorganized cell mass i.e. callus through organogenesis or embryogenesis |
| Embryo formation | Zygotic embryo develops from fusion of male and female gametic cells | Somatice embryo develops from somatic (non-sexual) cells |
| Protective cover | It poses a natural seed coat for protection from external injuries and adverse environment | In place of natural seed coat (testa), somatic embryo poses an artificial gel matrix |
| Food materials | Reserve food for zygotic embryos is either endosperm or cotyledons | Encapsulating matrix serves as reserve food for somatic embryos in synthetic seed |
| Genetical status | It is not genetically identical (particularly cross-pollinated plants) | It may produce genetically identical to parent, if somaclonal variation is ignored |
| Polarity | Bipolar possessing rudimentary root and shoot | It is also bipolar, under suitable culture medium root and shoot initiation take place simultaneously |
| Dormancy | Zygotic embryos after forming embryonal axis with the apical meristem, prepare for dormancy | Somatic embryos do not undergo desiccation and dormancy and embryogenesis goes on from initial cell to plantlet without interruption |

### 5.1.3. Androgenic Embryos

#### *5.1.3.1. Anther Culture*

Androgenic callus is very interesting in tissue culture. A number of ploidy levels of cells can be obtained through anther culture. In general, anther culture is done to initiate ***haploid*** development. The cells of haploid plants contain single set of chromosomes. The phenotype of haploids is the expression of information which is contained in a single copy of genes either recessive or dominant. The chromosome complement of haploid can be doubled either by colchicine treatment or simply regeneration of plantlets from callus doubling the chromosome complement is some species. In barley and rice, there is a high percentage of double haploid or dihaploidregenerants (up to 90%, depending on genotype) owing to a single ***autoendroreduplication*** of the genome during the first division of the microspore. Accordingly, there is no need for an application of agent (e.g. colchicine) to induce chromosome doubling. Haploid production from mature pollen grains of *Ginko biloba* was initiated by Tulecke (1953). He could obtain haploid cells, but failed to regenerate. It was Guha and Maheswari (1964) of Delhi University for the first time obtained haploid plants by culturing anther of *Datura innoxia*. This success of haploid production using gametic cell opened up new vista in the field of haploid breeding, subsequently production of ***double haploid*** and ***dihaploid***. After Guha and Maheswari, it was Bourgin and Nitsch (1967) who obtained androgenic plants in *Nicotiana tabacum*. Using this haploid production technology, China released a number of anther/pollen derived varieties of rice for commercial cultivation.

The success of haploid production from anther culture depends on various factors as mentioned below:

a) Appropriate stage of pollen in the anther.

b) Anther wall.

c) Physical status of donor plant which may influence by weather parameters, photoperiodism and mineral nutrition.

d) Pretreatment of anthers, such as, cold treatment of anthers at 3-10 °C for 2-30 days (4 °C for 8-10 days for rice, Roy and Mandal, 2004a & 2005).

e) Medium composition of anther culture and type of culture (Liquid or semi-solid culture).

f) Incubation temperature.

The optimum stage is not uniform. It depends on species to species. The appropriate stage for anther culture may be just before or just after the first pollen mitosis to early binucleate stage (see Fig. 5.3). In cereals, the best stage appears to be the early or mid-uninucleate stage, i.e., before the first pollen mitosis. In contrast, in species like tomato and *Arabidopsis thaliana*, the optimum stage is when the pollen mother cells are in meiosis-I, while in *Brassica* trinucleate pollen grains (at the time of pollen shedding) are the best. Pollen grains of *Brassica* remain responsive throughout the maturation phase, but their auxin requirement increases with the pollen age. In tobacco, the beginning of starch accumulation in pollen grains marks the end of their embryogenic potential.

For rice anther culture, anther is collected from the primary panicles when it is still enclosed in leaf sheaths, that is, the distance between the auricle of the flag leaf and the first subtending leaf was ~6 cm. This distance was found to be optimal to obtain maximum microspores in the mid-uninucleate to early binucleate condition, the most responsive stage for anther culture (Roy and Mandal, 2004a, & 2005). Further, the presence of starch in early binucleate pollen is indicative of a lack of androgenic potential of the species, while in species having the potential starch is absent.

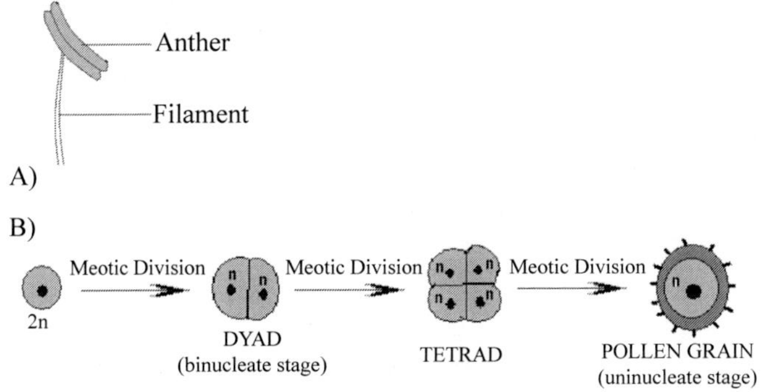


**Fig. 5.3.** Microsporogenesis in higher plants.
**A)** Androecium **B)** Different stages of microsporogenesis

The optimum concentration of medium components is the primary requirement for initiation of ***androgenesis***. In general, Nitsch and Nitsch (1969) medium is well know as anther culture. However, other media, such as, MS (with minor modification) and $N_6$ (Chu, 1978) are also used for culture of anthers of a large number of crops. Chinese scientist used N6 medium for anther culture of wide range of species including rice, wheat, rubber, etc. Sucrose in the anther culture medium plays an important role as a nutrient as well as an osmoprotectant. The sucrose content in the general tissue culture medium is 2-3%, whereas the anther culture medium contains higher percentage (5-6%) of sucrose. In some special cases, the percentage is still higher, for example, *Brassica* sp. requires 12-13% sucrose for androgensis. Higher percentage (6%) of sucrose in cereal anther culture increases the frequency of green plantlets. Other sources of energy source are also used in cereal anther culture, *viz*. maltose (6%) was found to be most suitable for androgenic callus development and subsequently regenerate good number of green plantlets (Roy and Mandal 2004a).

In several plants regeneration of artificial seeds into plantlets has been reported. However, information about production of artificial seeds from androgenic embryos derived from androgenic callus is scanty. Haploid plant breeding has been found to be well established in many crops (Aljera *et al.*, 1995; Palmer *et al.*, 1996; Roy and Mandal, 2004a & 2005). The induction of pollen embryogenesis, which genetically differs from zygotic embryogenesis, may be used for synthetic seed production. In rice, Roy and Mandal (2006) developed protocol for rapid and recurrent mass multiplication of androgenic embryos, pro-embryo and embryo-like-structure of IR 72 an elite *indica* cultivar. These embryos and embryo-like-structures can be used as source of synthetic seed production Fig. 5.4.

**Fig. 5.4.** Anther culture and plantlet regeneration of rice. **(a)** Plated anthers showing androgenic callus induction on N6 medium supplemented with 15 % coconut water and 6% sucrose; **(b)** Stereomicroscopic view of emerging androgenic calli by rupturing the anther wall; **(d)** Regeneration of albino plantlets on MS containing 1 mg/L kinetin + 1 mg/L BAP and 0.5 mg/L NAA; **(e)** Regeneration of green plantlets on MS containing 1 mg/L kinetin + 1 mg/L BAP and 0.5 mg/L NAA; **(f)** Fertile plant developed from anther culture.

Serious bottleneck in AC response, particularly in *indica* rice; reduced regeneration, low number of DHs recovery, reduced regeneration of green plantlets, abundance of ***albino*** and high frequency of haploids in primary regenerants. Albinism is a limiting factor of androgenic haploid of all cereals crops. Gynogenetic path of haploid development may solve the problem of albinism in these crops.

### Pretreatment of flower buds/inflorescences

Exposure of excised flower buds to a low temperature at 3-5 °C for 2 days or at 7-8 °C for 12 days prior to removal of anthers for culture may markedly enhance the recovery of haploid plants. In some species, however, a brief exposure of anthers to a high temperature is reported to have an enhancing effect, e.g., at 35 °C for 24 h in *Brassica campestris*. It contrast, the best pretreatment for cereals like wheat, rice, barley, etc. seems to be 3-28 days at 4-10 °C; this increases the frequency of green plants. The pretreatment temperature and duration may be considerably affected by plant species, genotype and stage of anther development.

The rice panicles along with the boot leaf sheath are washed thoroughly in tap water and spread with 70% ethanol. They are covered with moist tissue paper, kept in polyethylene bags and cold shocked for 8 days at 8 °C in a BOD incubator prior to anther plating.

### Inoculation of anthers of rice

- At the day of culture, selected spikelets/ flower buds are surface sterilized in tissue culture bottles (300 ml capacity) with 0.1% or 0.2% freshly prepared $HgCl_2$ solution for 10 minutes.
- The $HgCl_2$ was then drained off and the panicles are washed four times in sterile distilled water. The bottles were kept inverted on sterile tissue paper to drain out excess water.
- Fifty to sixty spikelets of rice are cut at a time on sterile petridishes under laminar airflow bench. When the spikelet surface becomes dried, individual spikelets were cut at the base to free the anthers from the filaments. In case of flower buds, remove the sepals and petals, and aseptically separate the anthers from the buds.
- They were plated aseptically onto 60 × 15 mm diameter radiation-sterile Petri-dishes containing suitable callus induction medium.
- The cultures were sealed with parafilm.
- Then the inoculated cultures are kept in dark at 25 ± 2 °C.

- The plates were under examination periodically at weekly interval to observe the progress in respect of callus formation.

### *5.1.3.2. Microspore Culture*

Haploid plants can also be regenerated from microspore culture. Anther culture may associate with production of diploid plantlets from anther wall or from other parts of anther other than pollen. Thus anther derived plantlets are with various ploidy levels. This can be avoided by culturing isolated pollen. Microspore culture usually produces homogenous population, whereas anther culture could constitute a heterogenous population. Tulecke (1953) was able to obtain callus from isolated microspore culture *Ginkobiloba.* The first microspore culture was reported by Kameya and Hinata (1970) of *Brassica oleracea* and the hybrid *B. alboglaba*×*B. oleracea.*

Takahata *et al.* (1992) produced many embryos when isolated microspores of Chinese cabbage, broccoli and rape were cultured *in vitro*. Tolerance of desiccation of embryos at the cotyledon stage was highest with the application of 10-100 mM ABA and the conversion frequency was 40-50%. Viability was maintained after desiccation for 6 months at room temperature without humidity control. Desiccated embryos lost their viability if not treated with ABA.

- **Isolation and inoculation of pollen for callus induction**
  1. Do not collect the flower-buds until the laboratory is ready. Appropriate sized and stage of buds is collected.
  2. Pretreatment may be given if needed. Usually cold pretreated at 4-10 °C for 3-15 days depending on the species.
  3. Surface-sterilize the unopened flower buds with suitable surface sterilizing agent.
  4. Wash thoroughly with sterile-distilled water to remove the surface sterilizing agent from the surface of the flower buds.
  5. Place about 40-60 anthers in a sterilized beaker containing 20 ml liquid anther culture medium and press the anthers by means of a glass-rod to squeeze out the microspores.
  6. The solution is filtered through a nylon sieve.
  7. Centrifuge the filtered solution.
  8. Remove the supernatant and resuspend the pallet containing pollen in fresh medium. Repeat the same for 2-3 times.
  9. Finally transfer the pollen to callus induction medium.
  10. Incubate the culture plate in culture room under 500 lux light and 25 ± 2 °C.

### *5.1.3.3. Ovule Culture*

Haploid plants have been successfully developed from culture of female gametophytic cells, that is, the egg nucleus or ovum. It was considered as an alternative means of haploidy as well as the expression of totipotency of female gametophytic cells in ***angiospermic*** plants. The first gynogenic haploid was obtained in barley (San Noeum, 1976) culturing ovaries. Subsequently, Chinese scientists could regenerate haploid plants in rice, wheat, sunflower, sugar beet and onion by culturing female gametophytic cells. Haploid have also been obtained from unpollinated ovule culture of *Gerbera jasmesonii, Liliumdavidii, Zea mays* etc.

The appropriate time of embryo sac for its culture is uninucleate to mature stage. But it may differ species to species, such as nearly mature (1-2 days before anthesis), embryo sac mother cell to megaspore tetrad stages are also suitable for culture. The explant for development of gametophytic haploid may be the ***ovary***, isolated ovule or even unhusked flower (Fig. 5.5).

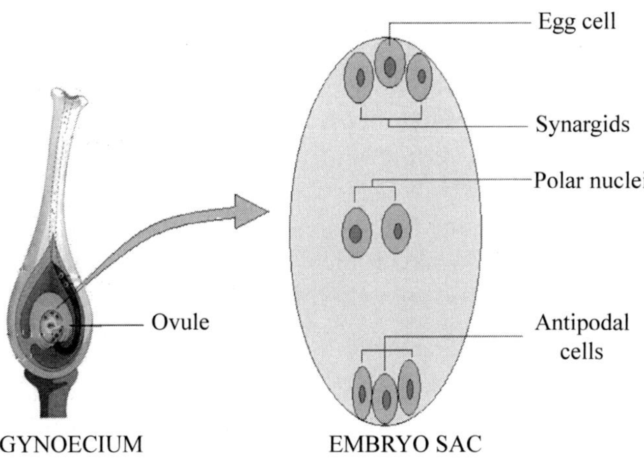


**Fig. 5.5.** A sectional view of ovary and detail of embryo sac

## 5.1.4. Recurrent Embryogenesis

The genetic improvement of crop plants *en route* biotechnological approaches largely depends upon the maintenance of differentiated cultures of callus or redifferentiated embryos. Subsequently the success of synthetic seed production depends on how best callus develop and plantlet regeneration achieved and continuous supply of embryogenic calli. Long-term maintenance of embryogenic masses in culture tubes or mechanically stirred bioreactors requires frequent transfer of tissue to fresh media, which is both labour intensive and costly. However, over time the morphogenic competence of differentiated cultures declines (Lynch and Benson, 1991). Therefore, new culture has to be regularly initiated and characterized in order to maintain a constant supply of embryogenic callus. This approach is highly cumbersome.

#### *5.1.4.1. Repeated Subculture*

To cope with these difficulties, recurrent embryogenesis can be practiced. It is also known as repetitive, accessory, proliferative or secondary embryogenesis. On the other hand, this can be referred as ***embryo cloning*** technique. The power of embryo cloning technique can be used in synthetic seed preparation for uninterrupted supply of somatic embryos. The maintenance of recurrent cycle of somatic embryogenesis can be spontaneous as is the case with alfalfa, *Medicago sativa* (Lupotto, 1983). The cycles were maintained without growth regulators (Lupotto, 1986), or with specific growth regulator at specific concentration (Roy and Mandal, 2006). More recently, however, the initiation of recurrent culture requires that the developing embryos can be locked into a developmental stage beyond which they cannot proceed, thereby repeating a cycle. This can be achieved through its initial exposure to a very high auxin concentration such as 40 mg/L 2,4-D followed by maintenance of the recurrent system using a lower level of auxin, such as 5 mg/L of 2,4-D (Finner and Nagasawa, 1988), which prevent the transition from pro-embryogenic to embryogenic development. Onay *et al.* (1996) have reported that the encapsulated embryogenic masses recovered their original proliferate capacity after two month storage following two subcultures.

Roy and Mandal (2006) developed a protocol for rapid and recurrent mass-multiplication of androgenic embryos and microtillers of indica rice var. IR 72. Fig. 5.6 represents the flow diagram of recurrent multiplication of embryos, pro-embryos and microtiller of rice var. IR 72. Those embryo masses and microtillers were used to prepare synthetic seeds (Roy and Mandal, 2008) and germinated *in vitro* and *in vivo* on vermiculite. Petiard *et al.* (1993) elaborated tool for large-scale propagation of coffee (*Coffee canephora*) and carrot from mass-somatic embryogenesis in bioreactors. Subsequently they have stressed upon the economics of mass somatic embryogenesis.

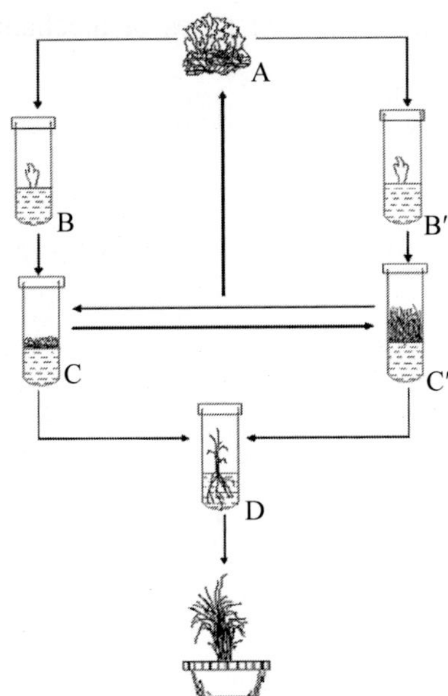


**Fig. 5.6.** Flow diagram of recurrent-mass-multiplication of embryos, pro-embryos and micritillers of rice (Roy and Mandal, 2006). **A)** Dormant embryos and pro-embryos on MS medium with 6 mg/L of BAP; **B)** Single dormant embryo/pro-embryo cultured on MS medium with 6 mg/L of BAP; **C)** Mass multiplied dormant embryos and pro-embryos on medium fortified with BAP 6 mg/L; **B′)** Single dormant embryo/pro-embryo cultured on MS medium fortified with 4 mg/L of kinetin; **C′)** Mass multiplied microtillers on MS medium fortified with kinetin 4 mg/L; **D)** Plantlet developed from embryos/pro-embryos/microtillers on MS medium fortified with 1 mg/L of IAA or IBA; **E)** Fertile plant develop *in vivo* from embryos/proembryos/ microtillers.

### *5.1.4.2. Continuous Suspension Culture*

In a continuous culture, the cell population is maintained in a steady state by regularly replacing a portion of the used or spent medium by fresh medium. Such culture systems are of either (1) closed or (2) open type. In a closed continuous culture, cells are separated from the used medium taken out for replacement, and added back to the culture so that cell biomass keeps on increasing. In contrast, both cells and the used medium are taken out from open continuous cultures and replaced by equal volume of fresh medium.

The replacement volume is so adjusted that cultures remain at submaximal growth indefinitely. The open cultures are of either turbidostat or chemostat types. In a **turbidostat**, cells are allowed to grow up to a preselected turbidity (usually measured as OD) when a predetermined volume of the culture is replaced by fresh normal culture medium.

But in a **chemostat**, a chosen nutrient is kept in a concentration so that it is depleted very rapidly to become growth limiting, while other nutrients are still in concentrations higher than required. In such a situation, any addition of the growth limiting nutrient is reflected in cell growth. Chemostats are ideal for the determination of effects of individual nutrients on cell growth and metabolism.

### 5.1.5. Embryo Maturation and Germination

Low auxin level is required for formation of shoot meristem. Alternately high auxin content inhibits the development and growth of the shoot meristem, if young pro-embryos are not transfered to a low-auxin or zero-auxin medium after its induction (Parrott *et al.*, 1988). Activated charcoal may be added to the medium to remove the activity of excess auxin from the somatic embryos (Buchheim*et al.*, 1989).

Sucrose is the carbon source in the artificial medium, which promotes production of vigourous plantlets. The sucrose level varies as 3-6%, although progressively increasing levels up to 40% have been utilized for some species (Buchheim*et al.*, 1989). High concentration of sucrose in the medium generally cause desiccation in the embryos, however, for some species, efficient conversion to plantlets also requires the imposition of temporary desiccation before germination. This procedure, which mimics seed maturation *in vitro* may be necessary to trigger metabolic process required for germination and seedling growth (Rosenberg and Rinne, 1988). Carman (1989) reported that gradual reduction in somatic potential through desiccation of mature somatic embryos of wheat, which improved germination.

## 5.2. NON-EMBRYOGENIC SOURCES OF PROPAGULES

Conversion is the most important aspect of the success of the synthetic seed technology. In contrast to somatic embryo, which is a bipolar structure, whereas shoots, microtilers, microbulbs, rhizomes, shoot buds etc. are unipolar structure and do not posses root meristems. In case non-embryogenic culture, organs are not induced to form callus, the existing growing tip(s) directly grow on a chemically defined medium. Therefore, root induction is essential for the success of synthetic seed preparation using non-embryogenic explant. Different non-embryogenic explants for preparation of synthetic seeds are being explained briefly in this chapter.

### 5.2.1. Axillary and Apical Bud Culture

Axillary shoot buds and apical shoot tips are suitable for encapsulation studies of artificial seeds as they possess great potential for plant development from preexisting meristematic tissue. For production of synthetic seeds from axillary shoot buds (Fig. 5.7) and apical shoots, theses organs are usually first treated with auxins for root induction and then their microcuttings are encapsulated.

Bapat and Rao (1990) and Ganapathi *et al.* (1992) have described how encapsulated buds of banana and mulberry were converted into plantlets without specific root induction treatments. In case of banana, the synthetic seed technology may improve the quality of plantlets and reduce the cost of plantlets (suckers) production. This technology was successfully employed in banana (Ganapathi *et al.*, 1992, 2001; Matsumoto *et al.*, 1995a) and they obtained high percentage of *in vitro* conversion.

In contrary, root induction treatments before encapsulation of olive (cv. Moraliolo) buds resulted in low percentage of plantlet regeneration, which was in any case obtained only from apical microcuttings (Micheli *et al.*, 1998). Modification of IBA concentrations in the root induction phase and/or the composition of the nutrient enriched capsules and/or the sowing medium increased plantlet regeneration rate by up to 30%. Many authors followed encapsulation using these uninodal explants for efficient shoot regeneration through direct organogenesis (Andriani *et al.*, 2000; Mandal *et al.*, 2000; Sicurani *et al.*, 2001; Alvarez *et al.*, 2002; Brischia *et al.*, 2002; Priya *et al.*, 2003). Mandal *et al.* (2001) encapsulated basils (*Ocimum* spp.) in calcium alginate gel to produce synthetic seeds. Complete plantlets were retrieved from encapsulated nodal segments of *O. americum, O. basilicum* and *O. sanctum.* Capuano *et al.* (1998) reported that conversion of synthetic seeds obtained with axillary microcuttings of M.26 apple rootstock always occurred at a very low rate (only 25%) following 6 days of root ***primodial*** initiation culture and cold storage. In contrast, apical microcutting, reached 85% conversion with a 24.6 μM IBA treatment and 3 days root promodial initiation culture without cold storage. The results encourage the use of encapsulated unipolar explants for synthetic seed technology. This kind of capsule could be useful in exchange of sterile material between laboratories due to small size and relative ease in handling these structures, or in germplasm conservation with proper preservation techniques (Piccioni and Standardi, 1995), or even in plant propagation and nurseries.

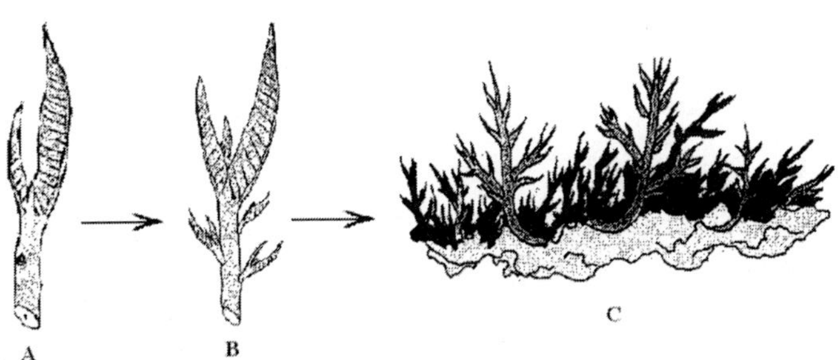


**Fig. 5.7.** Axillary shoot tip proliferation. **A)** Axillary bud suitable for culture; **B)** Initiation of proliferation of axillary bud; **C)** Proliferated axillary buds on synthetic medium suitable for *in vitro* encapsulation.

### 5.2.2. Meristem Culture

The cultivation of axillary or apical meristem (Fig. 5.8), involves in the development of an already existing shoot apical meristem and the regeneration of adventitious roots. Meristem culture may be utilization for rapid multiplication and clonal propagation of adventitious shoot, bud regeneration, for germplasm conservation and/or for synthetic seed technology. When a meristem is cultured *in vitro*, it produces a small plant bearing 5/6 leaves. This could be obtained within a few weeks. The stem is cut into 5/6 small microcuttings, which under favourable conditions develop into fully-grown plants. Since these unipolar structures do not have root meristems, they should be inoculated to regenerate roots before encapsulation.

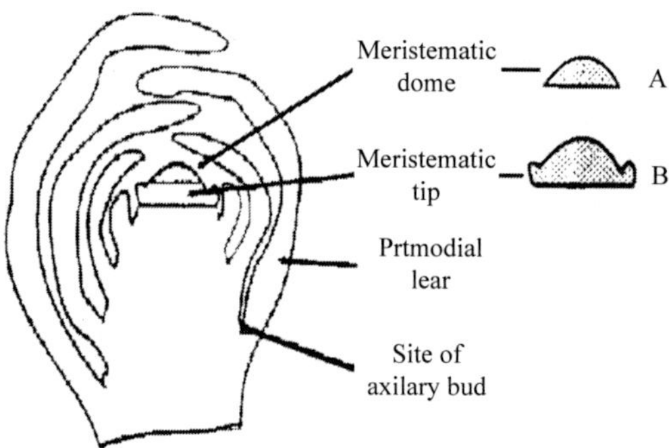


**Fig. 5.8.** Different parts of shoot tip, which can be used as explant for synthetic seed preparation. **A)** Meristem dome suitable for encapsulation; **B)** Meristematic tip suitable for encapsulation.

Meristem culture has advantage in production of disease free (particularly virus free) plants. Apical meristem in the infected plants are generally either free or carry a very low concentration of the viruses. The virus density increases in the older tissues corresponding to the increase in the distance from the meristematic tip because of the following reasons:

- Virus moves in the plant body through vascular system which is absent in meristem.
- Cells of the meristem divide rapidly, which does not allow the virus to settle in the cell.
- Meristem possesses high metabolic activity.
- Endogenous concentration of auxin in meristem is very high, which inhibit virus multiplication.

### 5.2.3. Protocorm or Protocorm-like Bodies

Since orchid seeds are lack of endosperm as well as hard seed coat and their embryo were made of undifferentiated cells, an artificial nutrient supply is necessary to nurish embryo during germination when they are grown asymbiotically. Due to all these inhibitory factors, their germination is almost negligible in natural condition. Therefore, artificial nutrient supply and protective seed coat may be provided around the orchid embryos by encapsulating it with a suitable encapsulating matrix. In orchids, different workers have produced synthetic seeds by encapsulating the protocorm or protocorm-like bodies in sodium alginate gel viz. Bhattacharjee *et al.* (1998) in *Phalaenopsis*, Datta *et al.*,(1999) in *Geodorum densifolium*, Corrie and Tandon (1993) in *Cymbidium giganteum*, Malemngaba *et al.* (1996) in *Phaius tonkervillae*, Sharma *et al.* (1992) in *Dendrobium wardianum*, and *Rhyncostylis*. Corrie and Tandon (1993), *Aridesodoratum*, *P. takervillae* have reported that the encapsulated protocorms of *Cymbidium giganteum* gave rise to health plantlets upon transferring to nutrient medium or directly to sterile sand and soil. They found that conversion frequency was high in both *in vitro* (100%) and *in vivo* (85% in sand, 64% in sand and soil mixture) conditions. Bhattacharjee *et al.* (1998) encapsulated protocorms of $F_1$ hybrid of *Phalaenopsis* cv. Tane and *Phalaenopsis* cv. Abrae. The encapsulated seeds were stored at 4 and 25 °C. All synthetic seeds developed into young plants. However, germination rate declined as storage duration was extended and germination was faster following storage at 4 °C than at 25 °C. Datta *et al.* (1999) also encapsulated 30 day-old protocorm-like bodies in sodium alginate and they found 88% germination of *G. densifolium*. They observed that about 85% of encapsulated protocorms stored in room temperature in plates regenerated within 35-40 days in all the four species, while that of encapsulated protocormsoverlaid with liquid paraffin and stored at 4 °C for 200 days showed 70% viability.

### 5.2.4. Young Seed Embryo

In embryo culture, young embryos are removed from mature/developing seeds and are placed on a suitable nutrient medium to obtain synthetic seeds (Fig. 5.9). The viability of recalcitrant seeds is very limited, 7-20 days of harvest under natural condition. The production of synthetic seeds from seed embryo may be used for long-term storage of recalcitrant seeds (Sudhakar *et al.*, 2000; Sunilkumar *et al.*, 2000). This type of synthetic seeds contains plant propagules as bipolar embryos. Simultaneously the root and shoot development may be observed on germination.

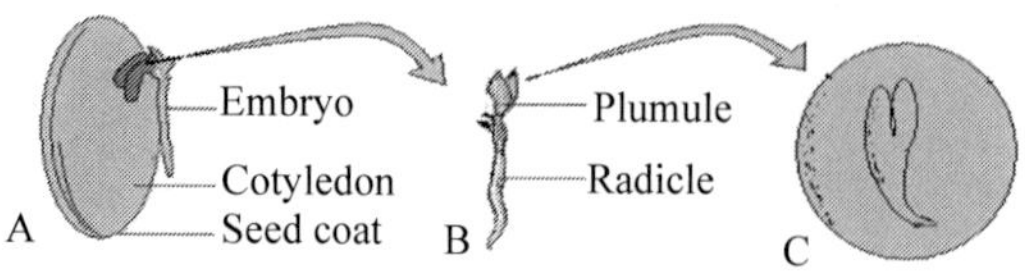


**Fig. 5.9.** Matured seed embryo as source of synthetic seed production. **A)** Dicot seed; **B)** Seed embryo and **C)** Encapsulated matured seed embryo.

### 5.2.5. Mature Seeds

Small seeds also can be used to prepare synthetic seed. It is difficult to maintain the number of seed per pit and spacing for small seeds in mechanical sowing. The synthetic seed technology can be used to increase the size of small seeds suitable for mechanical sowing. This technology also can be used to preserve seeds under cryogenic condition.

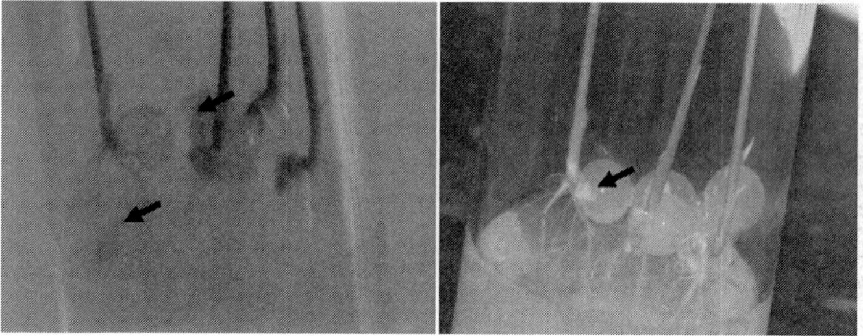

**Fig. 5.10.** Synthetic seed prepared from dehusked rice (arrows sowing the dehusked rice).

### 5.2.6. Bulblets and Microtillers

Microshoots (for dicot plants), microtillers (for monocot plants), and *in vitro* bulblets (microbulbs) can be encapsulated to prepare synthetic seeds. Andriani *et al.* (2000) encapsulated excised buds from *in vitro* proliferated shoots of the variety Hayward kiwifruit (*Actinidia deliciosa*) for preparation of synthetic seeds. *In vitro* bulblets of garlic were encapsulated and successfully regenerated on synthetic medium (Bekheet, 2006).

Roy and Mandal (2008) used androgenic microtillers of rice var. IR72 for encapsulation in sodium alginate (Fig. 5.11). Taha *et al.* (2012) prepared synthetic seeds from 3–5 mm microshoots of *Oryza sativa* L. Cv. MRQ 74. Microtillers/microshoots were encapsulated in 3% (w/v) sodium alginate, 3%s sucrose, 0.1 mg/L BAP, and 0.1 mg/L $\alpha$-Naphthalene acetic acid (NAA). Germination and plantlet regeneration of the encapsulated seeds were tested by culturing them on various germination media. The effect of storage period (15–30 days) was also investigated. The maximum germination and plantlet regeneration (100.0%) were recorded on MS media containing 3% sucrose and 0.8% agar with and without 0.1mg/L BAP. However, a low germination rate

(6.67%) was obtained using top soil as a sowing substrate. The germination rate of the encapsulated microshoots decreased from 93.33% to 3.33% after 30 days of storage at 4 °C in the dark.

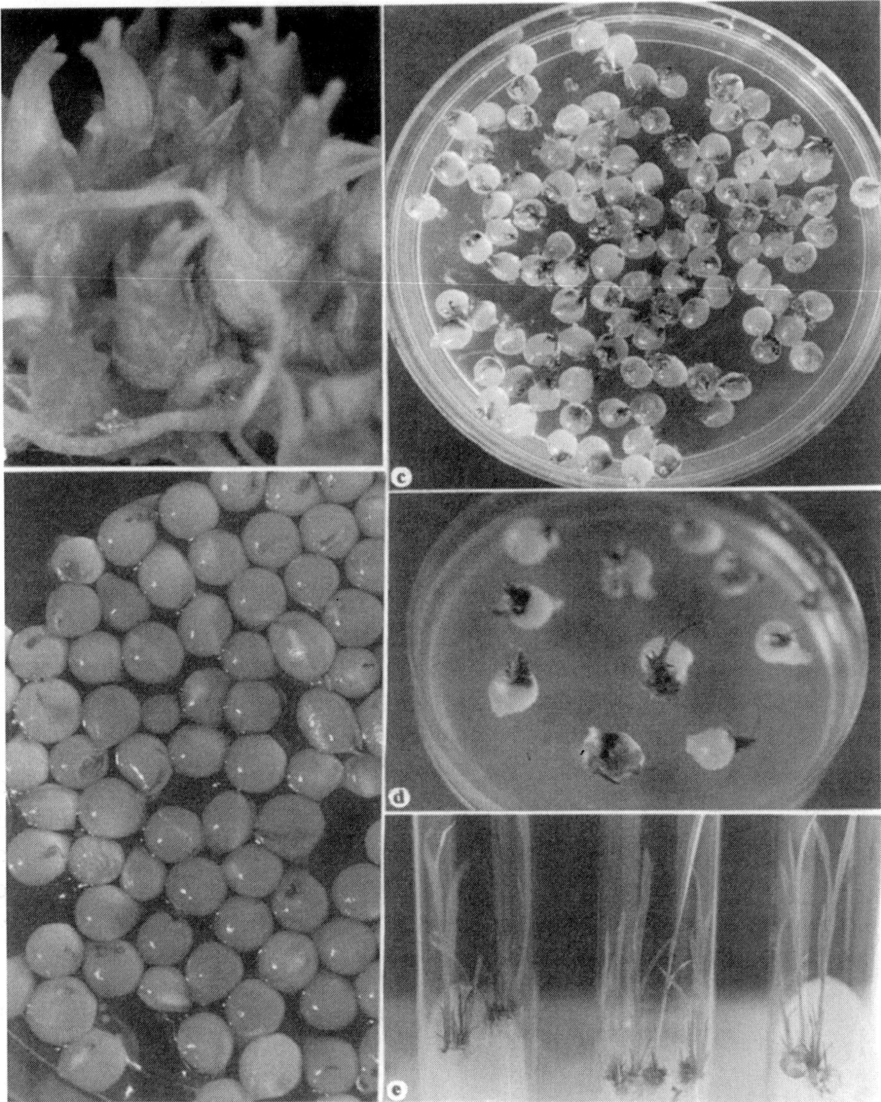

**Fig. 5.11.**Pictorial depiction of synthetic seed production and plantlet regeneration in rice (Roy and Mandal, 2008). **a)** Stereomicroscopic view of germinating microtillers; **b)** Encapsulated microtiller in sodium alginate (4%) beads; **c)** Mass germination of beaded microtillers on MS medium with no hormones; **d)** Seedling elongation from germinating synthetic seeds; **e)** Growing plantlets under *in vitro* culture condition on MS basal medium.

## References

Aitken-Christie J, Kozai T, Smith MAL.1995. Glossary. In: Aitken-Christie J, Kozai T, Smith MAL (eds) Automation and Environmental Control in Plant Tissue Culture, Kluwer, Dordrecht. pp ix-xii.

Aljera MS, Zapata D, Khush GS, Datta SK. 1995. Utilization of anther culture as a breeding tool in rice improvement. In: Current Issues in Plant Molecular and Cellular Biology, Terzi M *et al.* (eds.), Kluwer Academic Publishers, The Netherlands, pp. 137-142.

Alvarez CR, Gardi T, Standardi A. 2002. Plantlets from encapsulated *in vitro* derived microcuttings of kiwi fruit (cv. Top star R). Agricultura Mediterranea. 132(3-4): 246-252.

Anandarajah K, McKersie BD. 1990b. Enhanced vigor of dry somatic embryos of *Medicago sativa* L. with increased sucrose. Plant Sci. 71: 261-266.

Andriani M, Piccioni E, Standardi A. 2000. Effect of different treatments on the conservation of "Hayward" Kiwifruit synthetic seeds to whole plants following encapsulation of *in vitro* derived buds. New Zealand J Crop Hort Sci. 28(1): 59-67.

Ara H, Jaiswal U, Jaiswal VS. 2000. Synthetic seed: Prospectus and limitations. Current Sci. 78(12): 1438-1444.

Bapat NA, Mahtre M, Rao PS. 1987. Propagation of *Morus* indica L. (Mulberry) by encapsulated shoot buds. Plant Cell Rep. 6: 393-395.

Bapat VA, Rao PS. 1990. *In vitro* growth of encapsulated axillary buds of mulberry (*Morisindica* L.). Plant Cell Tiss Org Cult. 20: 69-70.

Bekheet SA. 2006. A synthetic seed method through encapsulation of *in vitro* proliferated bulblets of garlic (*Allium sativum* L.).Arab J Biotech. 9(3): 415-426.

Bhattacharjee S, Khan HA, Reddy PV, Bhattacharjee S. 1998.*In vitro* seed germination, production of synthetic seeds and regeneration of plantlets of *Phalaenopsis* hybrid. Annals AgrilSci, Cairo. 43(2): 539-543.

Bourgin JP, Nitsch JP. 1967. Obtention de *Nicotiana*haploides á partird'étaminescultivées*in vitro*. Ann Physiol Veg. 9: 377-382.

Brischia R, Piccioni E, Standardi A. 2002. Micropropagation and synthetic seed in M.26 application root stock (II): A new protocol for production of encapsulated diffentiatingpropagules. Cell Tiss Org Cult. 68(2): 137-141.

Buchheim JA, Colburn SM, Rach JP. 1989. Maturation of soybean somatic embryos and transition to plantlet growth. Plant Physiol. 89: 768-775.

Capuano G, Piccioni E, Standardi A. 1998. Effect of different treatments on the conversion of M26 apple rootstock synthetic seeds obtained from encapsulated apical and axillary micropropagated buds. J. HorticSciBiotechnol.73: 299-305.

Carman JG. 1989. The *in ovulo* environment and its relevance to cloning wheat via somatic embryogenesis. *In Vitro* Cell Dev Biol. 25: 1155-1162.

Chee RP, Cantliffe DJ. 1992. Improved production procedures for somatic embryos of sweet potato for a synthetic seed system. Hort Science. 27(12): 1314-1316.

Chu CC. 1978. The $N_6$ medium and its applications to anther culture of cereal crops. In: Proceedings of Symposium on Plant Tissue Culture. Science Press, Perking. pp. 43-50.

Corrie S, Tandon P. 1993. Propagation of Cymbidium giganteum Wall, through high Frequency Conversion of Encapsulated Protocorms under *In Vivo* and *In Vitro* Conditions.Indian J ExpBiol. 31: 61-64.

Datta KB, Kanjilal B, Sarker D-de, De-Sarkar D. 1999. Artificial seed technology: Development of a protocol in *Geodorumdensiflorum* (Lam) Schltr. – an endangered orchid. Current Sci. 76(8): 1142-1145.

Finner JJ, Nagasawa A. 1988. Development of an embryogenic suspension culture of soybean (*Glycine max* Merrill.). Plant Cell Tiss Org Cult. 15: 125-136.

Flacinelli M, Piccioni E, Standardi A. 1993. Synthetic seed in crop plants: problems and prospects. SementiElette. 39(2): 3-13.

Ganapathi TR, Srinivas L, Suprasanna P, Bapat VA. 2001. Regeneration of plants from alginate-encapsulated somatic embryos of banana cv. 'rastali' (Musa spp. AAB group).*In vitro* Cell Dev Biol- Plant. 37: 178-181.

Ganapathi TR, Suprasanna P, Bapat VA, Rao PS. 1992. Propagation of banana through encapsulated shoot tips. Plant Cell Reports. 11(11): 571-575.

Ghosh B, Sen S. 1994. Plant regeneration from encapsulated embryos of *Asparagus cooperi* Baker.Plant Cell Rep. 13: 381-385.

Gray DJ, Purohit A. 1991. Somatic embryogenesis and development of synthetic seed technology. Critical Reviews Plant Sci. 10(1): 33-61.

Guha S, Maheswari SC. 1964. *In vitro* production of embryos from anthers of *Datura*. Nature, London.204: 497.

Ipekci Z, Gozukirmizi N. 2003. Direct somatic embryogenesis and synthetic seed production from *Paulownia elongate*. Plant Cell Rep. 22(1): 16-24.

Jain SM, Gupta PK, Netwon RJ. 1995. Somatic embryogenesis in woody plants, Kluwer Academic Publishers, Dordrecht.

Kamada H, Kobayashi K, Kiyosue T, Harada H. 1989. Stress induced somatic embryogenesis in carrot and its application to synthetic seed production. *In vitro* Cellular Dev Biol. 25(12): 1163-1166.

Kameya T, Hinata K. 1970. Induction of haploid plants from pollen grains of *Brassica*. Japan J Breed. 20: 82-87.

Kim YH, Janick J. 1990. Synthetic seed technology: improving desiccation tolerance of somatic embryos of celery. Acta Horticulturae. 280: 23-28.

Lai FM, McKersie BD. 1994a. Regulation of starch and protein accumulation in alfalfa (*Medicago sativa* L.) somatic embryos. Plant Sci. 100: 211-219.

Lai FM, McKersie BD. 1994b. Regulation of storage protein synthesis by nitrogen and sulpher nutrients in alfalfa (*Medicago sativa* L.) somatic embryos. Plant Sci. 103: 209-221.

Lulsdorf MM, Tautorus TE, Kikcio SI, Bethune TD, Dunstan DI. 1993. Germination of encapsulated embryos of interior spruce (*Piceaglaucaengelmannii* complex) and black spruce (*Piceamariana* Mill.). Plant Cell Reports. 12(7-8): 385-389.

Lupotto E. 1983. Propagation of an embryogenic culture of *Medicago sativa* L. Pflazenphysiol. 111: 95-104.

Lupotto E. 1986. The use of single somatic embryo culture in propagating and regulating lucern (*Medicago sativa* L.). Ann Bot. 57: 19-24.

Lynch PT, Benson EE. 1991. Cryopreservation, a method for maintaining the plant regeneration capacity of rice, cell suspension culture. In: Proceedings of the Second Rice Genetic Symposium, International Rice Research Institute, Los Banos, Philippines. pp. 321-332.

Malenganba H, Roy BK, Bhattacharya S, Deka PC. 1996. Regulation of encapsulated protocorms of *Phaiustankervilliae* stored at low temperature. Indian J Exp Biol. 34: 802-805.

Mandal J, Patnaik S, Chand PK. 2000. Alginate encapsulation of axillary buds of *Ocimumamericunum* L. (hoary basil), *O. basilicum* L. (sweet basil), *O. gratissium* L. (sharubby basil) and *O. sanctum* L. (sacred basil). In vitro Cellular Devl. Biol. Plant. 36 (4): 487-292.

Mandal J, Patnaik S, Chand PK. 2001. Growth response of the alginate-encapsulated nodal segments of *in vitro* cultured basils. J Herbs Spices Medicinal Plant. 8(4): 65-74.

Mathur J, Ahuja PS, Lal N, Mathur AK. 1989. Propagation of *Valerianawallichii* DC. using encapsulated apical and axial shoot buds. Plant Sci. 60(1): 111-116.

Matsumoto K, Matsumoto K, Hirao C, Teixeira J. 1995a. *In vitro* growth of encapsulated shoot tips in banana (*Musa* sp.). Acta Horticulturae. 370: 13-20.

McKersie BD, Senaratna T, Bowley SR, Brown DCW, Krochko JE, Bewley JD. 1989. Application of artificial seed technology in the production of hybrid alfalfa (*Medicago sativa* L.). *In Vitro* Cell Dev Biol. 25: 1183-1188.

Micheli M, Mencuccini M, Standardi A. 1998. Encapsulation of *in vitro* proliferated buds of olive.AdvHort Sci. 12(4): 163-168.

Nitsch JP, Nitsch C. 1969. Haploid plant from pollen grains.Science. 163: 85-87.

Onay A, Jeffree CE, Yeoman MM. 1996. Plant regeneration from encapsulated embryoids and an embryogenic mass of pistachio (*Pistacivera* L.). Plant Cell Rep. 15: 723-726.

Padmaja G, Reddy LR, Reddy GM. 1995. Plant regeneration from synthetic seeds of groundnut (*Arachishypogaea* L.). Indian J Expt Biol. 33: 967-971.

Palmer CE, Keller WA, Armison PG. 1996.Utilization of *Brassica haploids*. In : *In vitro* Haploid Production in Higher Plants. Vol. 3: Important selected plants, Jain SM, SoporyVeilleux RE (eds.). pp. 173-192.

Parrot WA, Dryden G, Voght S, Hildebrand DF, Collins GB, Williams EG. 1988. Optimization of somatic embryogenesis and embryo germination in soybean. *In Vitro* Cell Dev Biol. 24: 817-820.

Petiard V, Ducos JP, Florin B, Lecouteux C, Tessereau H, Zamarripa A. 1993. Mass somatic embryogenesis: a possible tool for large-scale propagation of selected plants. In: Proceedings of the Fourth International Workshop on Seeds: Basic and Applied Aspects of Seed Biology. Agers, France, 20-24 July, 1992. pp. 175-191.

Piccioni E, Standardi A. 1995. Encapsulation of micropropagated buds of six woody species. Plant Tiss Org Cult. 42(3):221-226.

Priya BT, ArumugamShakila, Shakila A. 2003. Synthetic seed production in banana :Adv Plant Sci. 16(1): 219-222.

Redenbaugh K, Paasch BD, Nichol JW, Kossler ME, Viss PR, Walkee KA. 1986. Somatic seed: encapsulation of asexual plant embryos. Biotechnology. 4: 797-801.

Rosenberg LA, Rinne RW. 1988. Protein synthesis during natural and precocious soybean seed (*Glycine max* L. Merr.) maturation. Plant Physiol. 87: 474-478.

Roy B, Mandal AB. 2004a. Toward development of mapping population through anther culture and conventional recombination breeding for moleculer tagging of salt-tolerant gene/s involving IR 28 and Pokkali. In: 9th National Rice Biotechnology Network Meeting, New Delhi, India, pp. 183-185.

Roy B, Mandal AB. 2005. Anther culture response in *indica* rice and variation in major agronomic characters among the androclones of a scented cultivar, Karnal local. African J Biotechnol. 4(3): 235-240.

Roy B, Mandal AB. 2006. Rapid and recurrent mass-multiplication of androgenic embryos in *indica* rice.Indian J Biotechnol. 5(2): 239-242.

Roy B, Mandal AB. 2008. Development of synthetic seeds involving androgenic embryos and pro-embryos in an elite *indica* rice. Indian J Biotechnol. 7(4): 515-519.

San Noeum LH. 1976. Haploids D' *Hordiumvalgare* L. Par culture *in vitro* d' ovaries non fecondes. Ann Amelior Plantes. 26: 751-754.

Senaratna T, McKersie BD, Bowley SR. 1989. Desiccation tolerance of alfalfa (*Medicago sativa* L.) somatic embryos. Influence of abscisic acid, stress pretreatments and drying rate. Plant SciLimerick. 65(2): 253-259.

Senaratna T. 1992. Artificial Seeds. Protech Advance. 10(3): 397-392.

Senaratna T, McKersie BD, Bowley SR. 1990. Induction of desiccation tolerance in somatic embryos. *In vitro* Cellular Dev Biol. 26(1): 85-90.

Sharma A, Tandon P, Kumar 1992. Indian J. Exp Biol. 30:747-748

Sicurani M, Piccioni E, Standardi A. 2001. Micro-propagation preparation of synthetic seed in M.26 apple root stock I: attempts towards saving labour in the production of adventitious shoot tips suitable for encapsulation. Plant Cell Tiss Org Cult. 66(3): 207-216.

Sudhakar K, Nagraj BN, Santosh Kumar AV, Sunilkumar KK, Vijay Kumar NK. 2000. Seed Res. 28(2): 119-125.

Sunilkumar, KK, Sudhakar K, Vijaykumar NK. 2000. Seed Res. 28(2): 126-130.

Taha RM, Saleh A, Mahmad N, Hasbullah NA, Mohajer S. 2012. Germination and Plantlet Regeneration of Encapsulated Microshoots of Aromatic Rice (*Oryza sativa* L. Cv.MRQ 74). The Scientific World J. 2: 1-6.

Takahata Y, Wakui K, Kaizuma N, Brown DCW. 1992. A dry artificial seed system for *Brassica* crops. ActaHorticulturae. 319: 317-322.

Tulecke W. 1953. A tissue derived from the pollen of *Ginkobiloba*. Science, New York. 117: 599-600.

Xing XH, Shen YW, Gao MW, Yin DC, Xing XH, Shen VW, Gao MW, Yin DC. 1995. Studies on production of artificial seeds of rice hybrid between indica and japonica. ActaAgronomicaSinica. 21(1): 45-48.

# 6

# Uses of Artificial Seeds

In fact, as the concept of synthetic seed involves the use of small propagules and enables the direct sowing of this material *in vitro* or *in vivo*, the technology could provide great flexibility to the breeders by not only reducing the costs when large quantities of propagules are required for handling, shipping and planting, but also eliminating the acclimation step when direct sowing *in vivo* is applied (Onishi *et al.* 1994).

The artificial seed technology has been applied to a number of plant species belonging to angiosperms. The exact application of artificial seeds will vary from species to species. In self-pollinated crops that currently have good seed production systems, synthetic seeds will not have any practical applications, but in cross pollinating species, especially those where seed production is difficult and expensive, synthetic seeds offer many advantages and opportunities. The importance of encapsulation technique as an aid to propagation in various crop plants has been discussed by many workers. By combining the benefits of vegetative propagation system with the capability of long-term storage and with clonal multiplication, synthetic seeds have many diverse applications in agriculture (Gray and Purohit, 1991; Redenbaugh *et al.*, 1991; Redenbaugh, 1993). In some species seed propagation is not successful. The reason behind this may be due to heterozygosity of seed, small sized seed, presence of reduced or insufficient endosperm, and seedless varieties of crop species like grapes, watermelon, mango, guava etc. Sometimes association of mycorrhizal fungi is essential for seed germination, for example orchids. The germination of these seeds under natural conditions or under control environment is not possible. Propagating of these species through synthetic seed is the added advantage. The possible implications of synthetic seed technology have been detailed hereunder.

## 6.1. CLONAL PROPAGATION

Synthetic seed technology is an alternative to traditional micro-propagation for production and delivery of cloned plants. It offers the possibility of low cost, high-volume propagation system that will compete with true seeds and transplants. The explants used in synthetic technology are somatic in origin, which satisfy the asexual means of reproduction. However, the production of synthetic seeds by encapsulating vegetative parts such as shoot tips (Ganapathi *et al.*, 1992), axillary buds (Danso and Ford-Flyood, 2003), microtillers, microbulbs etc. had been used in recent years as more suitable alternative to somatic embryos. There are four distinct routes for clonal propagation through synthetic seeds:

1. Somatic embryo encapsulation
2. Adventitious shoot buds encapsulation
3. Encapsulation of adventitious modified structures like bulbils (in cardamom), bulblets (in lily), protocroms (in orchids), microtillers (in cereals) etc.
4. Axillary branching meristem culture. However, some variation may appear when somatic embryos are used for encapsulation.

The synthetic seed is currently considered as the most effective and efficient alternative technique for propagation of commercially important plant species that had problems in seed propagation and plants produced non-viable seeds or without seed. The implication of this technology in plant propagation has been individually briefly discussed below.

### 6.1.1. Production of true-to-type Plantlets

Artificial seeds have vast application in different fields of plant biotechnology for cultivation of various plant species. They offer the opportunity to store genetically heterozygous plants or plants with a single outstanding combination of genes that could not be maintained by conventional methods of seed production due to genetic recombination exists in every generation for seed multiplication (Gray, 1997).

The sexually reproduced seed in cross-pollinated crops is undesirable because it assures that the seeds are not alike genetically following meiotic recombination. Since source materials are somatic embryos, shoot tips and buds, the synthetic seed derived from the plant propagules are genetically similar to each other and the donor plant. *In vitro* plant propagation sometimes may have high level of somaclonal variation depending on genotype, explant source, *in vitro* period, and cultivation conditions in which culture is established.

Coniferous forest species can be propagated cheaply through seeds. The conventional breeding programs in these species are very time consuming because the life cycle of conifers is very long. Coniferous forests are very heterogeneous and that the seed of outstanding individuals will not necessarily give rise to improved offspring. Artificial seed has the ability to clone those overhanging trees at reasonable cost and in minimum time (Desai *et al.*, 1997).

Molecular characterization is a pre-requisite for identification of the *in vitro* produced plants. SDS-***PAGE*** protein analysis, ***RAPD*** and cytological studies in many published reports were used to proof the identity of regenerated plants *in vitro* to their intact true-to-type and also to asses the variations which may arise out of tissue culture and synthetic seed preparation process. The genetic stability of plants growing out after storage in encapsulated form was confirmed via protein SDS-PAGE (El-Kazzaz and Taha, 2002; Rady and Hanafy, 2004; Nower *et al.*, 2007). RAPD analysis was also done to establish the clonal fidelity of the plants derived from synthetic seeds of garlic (Bekheet, 2006), potato (Bordallo *et al.*, 2004). El-Halwagy *et al.* (2004) also did not detect any polymorphism in plantlets regenerated from frozen embryogenic axes of almond caused by cryopreservation using protein profiles, isozymes, RAPD analysis and cytological examination. Plants derived from artificial seeds of *Carica papayaihybrid* Yuan You 1 grew normally to flowering and set fruit. Ye *et al.*, (1993) found no variation among these plants in either morphological characters or isozymetic pattern of peroxidase, easterase, and amylase.

### 6.1.2. Fast Multiplication of Plantlets

The aim and scope for switching towards artificial seed technology was for the fact that the cost-effective mass propagation of elite plant genotypes will be promoted. An implementation of artificial seed technology to somatic embryogenesis or the regeneration of embryos is based on the vegetative tissues as an efficient technique that allows for mass propagation in a large scale production of selected genotype (Ara *et al.*, 2000). Production of true-to-type plant materials of tree species through traditional method, such as grafting, layering, budding etc. take long time for rooting and subsequent establishment. Rapid multiplication of plants on artificial media provides source materials for the production of synthetic seeds leading to continuous supply of plant materials in large quantities in a short time round the year.

### 6.1.3. Round the Year Production of Plant Propaguls

True seed production is seasonal. *In vitro* production of somatic embryos or production of microshoots is independent of season. These micropropagules can be used as for preparation of synthetic seeds to supply planting material round the year. One of the limitations of current micropropagation procedures is that the tissue culture facility and the greenhouse production facility must be physically linked; as well their production must be synchronized to meet peak market demands. Synthetic seeds could be a means to uncouple this linkage enabling the micropropagation to be done year round at a site distance from the production facility.

### 6.1.4. Overcome Self-Incompatibility

Artifical seed production is beneficial to improve planting efficiency of crops that are vegetatively propagated, such as, fruit and nut trees because of ***self-incompatibility*** and long breeding cycle. Similarly, conifers forest, such as pine and spruce trees, which are currently planted only by seed, would benefit from synthetic seed production because the long conifer life cycle of the trees are capable of bearing seed would be avoided.

The vast majority of fruit species are propagated by vegetative means because of the presence of self-incompatibility and breeding cycles very long. The use of artificial seed facilitates its spread. However, the most useful artificial seed would be in the conservation of germplasm of these species (Towill, 1988).

### 6.1.5. Propagation of Plants Associated with Problems through Seed Multiplication

Many important plants either do not produce viable seeds (thus being propagated only by cutting or grafting), or even if they can be propagated by seeds, several problems are faced, including inbreeding depression and non-homogeneity of cultivars. To avoid that, controlled breedings are often necessary, a hand labour practice which increases the price of the seeds. One example is the production of Cyclamens (Winkelmann *et al*. 2004).

Artificial seed technology can be very useful for the propagation of varieties of crop plants, specially crops for which true-seeds are not used or problem associated in seed multiplication (cucumber, geranium) and many vegetitavely propagated plants which are more prone to infections (day lily, garlic, potato, sugarcane, sweet potato, grape, and mango). In Orchids, the very small seed size and the requirement of an associatation with mycorrhizal fungi are great limitations to seed propagation.

Sowing seed of synthetic varieties is a common practice in forage species such as alfalfa (*Medicago sativa* L.) and orchargrass (*Dactylis glomerata* L.). Such varieties from selection and crossing of lines are phenotypically uniform but of different genotypes. These lines cross freely year after year to produce seeds and to originate heterozygous and heterogeneous populations. The use of artificial seed allows multiplication of outstanding genotypes and genetically uniform, since this method does not require that annually cross-pollination is carried out to produce plants (McKersie and Brown, 1997).

### 6.1.6. Propagation of Plant with Recalcitrant Seeds

The ***recalcitrant*** seeds cannot be dried. This is because the recalcitrant nature is supposed to be due to inherent structure of seed coat and storage tissue (King and Roberts, 1980). Therefore, long term storage of these species in seed banks is not easily possible. Alternative method for long term preservation of these plants species is the preparation of synthetic seeds. Production of synthetic seeds by using excised embryos is one of the recent methods by which the longevity of recalcitrant seeds can possibly be increased (Sunilkumar *et al.*, 2000; Sudhakar *et al.*, 2000). Synthetic seeds comprise of an embryo separated from its accessory structures and encapsulated in a hydrated gel capsule, which will permit natural development of embryo (Murashige, 1978).

### 6.1.7. Propagation of Seedless Fruits

Artificial seed production is an essential technique for the proliferation of plant species which are not able to produce seed, such as seedless grapes and seedless watermelon (Saiprasad, 2001; Cartes *et al.*, 2009). Triploid species produces no viable seeds for natural propagation. The triploid hybrids of many species are used commercially for production of seedless fruit or less seeded fruits. Every time seed of these triploids should be produced by crossing between tetraploid and diploid species. Use of synthetic seeds as propagules would be very advantageous. For example, synthetic seeds of seedless watermelon would produce very less number of true fertile seeds than the conventional varieties.

### 6.1.8. Propagation of Polyploids

Artificial seeds can be employed for the production of polyploids with elite traits, avoiding the genetic recombination when these plants are propagated using conventional plant breeding systems, thus saving on time and costs.

### 6.1.9. Propagation for Sterile Plants

Many plants are sterile and do not set seed. Somatic embryogenesis in many instances would be preferable to make cuttings as a means to propagate those plants. Somatic embryos embedded in synthetic gelling agents may act as seeds for propagation of those plants.

### 6.1.10. Propagation of Seasonal Plants

Natural seeds of many crop plants are seasonal and require specific conditions for storage and germination on sowing. Synthetic seeds can be produced round the year. Attree and Fourkey (1993) reported that a large number of quiescent somatic embryo of conifers could be produced throughout the year and stored for germination in the spring, which simplifies production system and provides plants of uniform size. One of the limitations of current micropropagation procedures is that the tissue culture facility and the green house production facility must be physically linked; as well their production must be synchronized to meet the market demands. Synthetic seeds could be a means to uncouple this linkage enabling the micropropagation to be done year round at a distance from the production facility.

### 6.1.11. Propagation of Plantshaving Dormant Seeds

Dormancy is in a general sense is the inactive period of seed embryo which has attained maturity up to when it begins to germinate. Seed dormancy in specific sense is the condition in a viable seed by which it prevents germination. It is a special mechanism of repression of all regeneration activities for germination of seeds. Dormancy is the common feature of natural seeds, but by means of artificial seeds the dormancy period can be reduced to a great extent, thereby shortening the life cycle of the crop.

### 6.1.12. Problems Associated with Natural Clonal Propagation

Some plants response better in*in vitro* clonal propagation than the natural clonal propagation. Pineapple guava (*Acca sellowiana*) showed better response to somatic embryogenesis and recalcitrant to conventional method of clonal propagation. Cangahuala-Inocente *et al.* (2007) have suggested that the use of synthetic seed was prerequisite for survival of plantlets in Pineapple guava.

### 6.1.13. Multiplication of Hybrids and Parental Lines

The synthetic seed system can be employed in the propagation of male and female sterile lines for production of hybrid seed. Many crops yield best when their seed is produced by hybridization of male and female lines (Saiprasad, 2001). This hybrid production is expensive, particularly when conducted manually, and yield of hybrid seeds are often low and hybridization success are not always assured. Synthetic seeds would allow the incorporation of specific new gene(s) into single outstanding hybrid that could then be produced asexually by somatic embryos. Brar *et al.* (1994) emphasized the need for research on artificial seeds in rice through somatic embryogenesis and outlined its impact on mass propagation of true-breeding hybrid. The application of synthetic seed for the production of hybrids in alfalfa has been discussed in detail by

McKersie *et al*., 1989; McKersie and Bowley, 1993. Sakamoto *et al*. (1991) also successfully used synthetic seeds of celery and lettuce hybrids. A similar system could be developed in other crop although the specific commercial applications will undoubtedly differ.

$F_1$ hybrid seeds of cucumber (*Cucumis sativa*) is produced in green house, thus production cost of the seed materials is high which is reflected in the prices of the seed and the final produce. Artificial seed technology based on somatic embryogenesis may be one solution to this problem (Pellinen *et al.,* 1997). Similarly, in the commercial sector, it is very difficult to produce low-cost hybrid seed species such as cotton (*Gossypiumhirsitum* L.) andsoybean (*Glycine max* Merril.) because they have cleistogamous flowers and abscission problems as the seed that is currently used comes from self-pollinating species. However, hybrid seed is produced in small quantities in a very laborious manner by hand pollination. This small volume of hybrid seed could be massively increased through artificial seed technology. Thus, the hybrid force would be used commercially to originate a significant reduction in costs (Tian and Brown, 2000).

### 6.1.14. Problem Associated with Symbiotic Germination in Orchid

Orchids occupy the top position among the flowering plants, in cut-flower production and as potted plants. In the natural environments the seedling orchids are found very rarely, usually at the bases of mature orchid plants. Germination percentage of orchid seed is very low on pots or on other media for plant cultivation. The seeds lack a metabolic machinery and functional endosperm and require a fungal association for germination. As the orchid industry is reluctant on micropropagation as a major source of planting material, orchid synthetic seeds are indispensable as they could be delivered easily like true seeds from commercial tissue culture laboratories to growers. This technique of synthetic seed production serves as a low-cost, high-volume propagation system (Saiprasad and Raghuveer, 2003). As seeds are very minute, Protocorm-Like Bodies (PLBs) are used for synthetic seed preparation of orchid this can obviate the routine high cost system of propagation. Since, PLBs can be directly sown on the soil by passing *in vitro* steps of micropropagation followed by hardening, synthetic seed system can revolutionize propagation and transportation of orchid (Singh, 2006).

### 6.1.15. To Overcome Acclimatization of Tissue Culture Derived Plants

Micropropagation needs acclimatization of tissue culture derived plants before they are transferred into field conditions (hardening) because of their tenderness to the absence of lignification and low cuticle formation. The synthetic seed techniques are associated with the elimination of the final micro-propagation stages- rooting and acclimatization, allowing direct delivery as well as the development of cost-effective propagation systems (Singh, 2006). The artificial seed coat in case of synthetic seed provides protection and may be used directly as true seeds.

### 6.1.16. To Overcome Long Juvenile Period

The crop plants having long juvenile periods, for example, citrus, mango, grapes, etc., the planting efficiency can be considerably improved by the use of synthetic seeds instead of cuttings of seed propagation (Naik and Chand, 2006).

### 6.1.17. Mass Multiplication of Transgenic and Elite Plants

Artificial seed production through the use of somatic embryos is an important technique for transgenic plants, where a single gene can be placed in a somatic cell and then this gene will be located in all the plants produced from this cell. Therefore, artificial seeds could be an efficient technology used for reproduction of transgenic plants (Daud *et al.*, 2008).

This newly emerging technology would also be useful for multiplication of genetically engineered plants (transgenic plants). ***Transgenic*** plants require separate growth facilities to maintain original genotypes may also be preserved using somatic embryos followed by preparation of synthetic seeds for multiplication or further propagation. Multiplication of elite plants selected in plant breeding programmes through somatic embryos avoids the genetic recombination, and therefore, does not warrant continued selection inherent in conventional breeding, saving considerable time and resources.

Biotechnology of forest trees possesses equal importance as that of field and horticultural crops. Development of transgenic inserting desirable gene(s) in forest trees is possible. The putative transformed tissue may be regenerated into plantlets. Regeneration of large numbers of transgenic plants is quite difficult or sometime impossible. The transformed plant can be used for clonal propagation through somatic embryogenesis followed by preparation of artificial seeds.

Crops from genetically modified plants have boomed in recent years. There is little information about what happens to these GMOs in the process of sexual reproduction. It is possible that during sexual multiplication, the introduced genes from other species are meiotically unstable and lost. With the use of artificial seed technology would avoid such risks. Similarly, this technology could be used in the propagation of somatic hybrids and cytoplasmic (obtained through protoplast fusion) and in sterile and unstable genotypes (Kumar, 1998).

### 6.1.18. Aforestation/ Reforestation

Forest tree species possesses great genetic variation than the cultivated herbaceous species. The size of the genome is quite larger for many trees, which is particularly important for genetic engineering. A further limitation in tree improvement using genetic engineering is that mature material must be collected from the field rather than the controlled environment. The problem of ecotypic of physiological variation from individual to individual is much greater due to the overall heterozygosity of most of the tree species. Thus, the desirable ecotype of the tree species can be clonally propagated through somatic embryogenesis followed by synthetic seed preparation.

The most important advantage of cloning by somatic embryogenesis followed by synthetic seed production is that embryogenic tissue can be cryopreserved without changing its genetic makeup and without loss of juvenility. This offers an opportunity to develop high value of clonal varieties in deforesting area for repropagating cryopreserved clones (Park *et al.,* 1998). The gel beads contain all the nutrients needed for the production of plantlets from the encapsulated propagules and also to protect the artificial seeds from microbial and insect attacks. The seeds are planted directly in soil where they give rise to healthy plantlets.

## 6.2. MECHANICAL PLANTING

Somatic embryogenesis and enhanced axillary bud proliferation system are the efficient techniques for rapid and large scale *in vitro* multiplication of elite and desirable plant species. Through these system a large number of somatic embryos shoot buds are produced which are used as efficient planting material as they are potent plant structures for plant regeneration either after having minor treatment with growth regulator or without any treatment. As naked propagules are sensitive to external environment, disease pathogens and desiccation, these should be planted only after hardening under glasshouse or nethouse condition. These *in vitro* derived propagulescannot be used for mechanical planting. To improve the success of mechanical planting/sowing of these propagules some protecting coating is essential. The synthetic seed production technology can provide protection for mechanical sowing directly in the field. The size and shape suitable for mechanical planting can be adjusted during synthetic seed production. Very small sized true-seed is difficult for mechanical sowing. These seeds can be embedded with the gelling agents to increase the size of individual seeds to fit for mechanical sowing.

## 6.3. REDUCES COST OF PRODUCTION

Regeneration from embryogenic callus to plantlets requires comparatively more culture medium than the volume of culture medium needs for encapsulation of somatic embryos, thus, reduces the cost of production. For example, one milliliter of culture medium is sufficient for encapsulation of a single shoot tip compared to 15-20 ml for conversion of shoot tip into plantlets through simple micro-propagation techniques.

Synthetic seeds of banana eliminate the two stage process of rooting and hardening of tissue cultured shootlets. Moreover, encapsulated shoot tips of banana is inexpensive, easier and safer material for germplasm exchange, maintenance and transportation.

## 6.4. PROTECTION FROM EXTERNAL ENVIRONMENT

A number of compounds namely, nutrients, growth regulators, anti-pathogen, insecticides, herbicides, bio-control agents, bio-fertilizers can be added with the gelling agents to improve the germination, conversion rate finally for the establishment of the synthetic seed-derived plants.

### 6.4.1. Protection from Insect Pests

In natural environment of forest many seedlings cannot convert into a well developed plant due to insect attacks at the early seedling stage. This problem has been detected in rauli-beech (*Nothofagus alpina*), a forest tree of Chile, insect attacks affect the development of this species (Burschel *et al.*, 1976; Cartes *et al.*, 2009). For this case, particular insecticide(s) can be added in the gelling matrix, which will prevent the insect attacks at the early seedling stages.

### 6.4.2. Protection from Diseases

Artificial seeds, which are produced using tissue culture techniques that are aseptic, are free of pathogens, giving great advantages to these materials for transport across frontiers and for avoiding the spread of plant diseases (Saiprasad, 2001; Nyende *et al.*, 2003, 2005). Artificial seeds are also valuable in terms of their role in providing protective coating, increasing the level of micropropagule success in the field. These micropropagules need a protective coating to increase successful establishment in the field situation because of the sensitivity of uncovered micropropagules to drought and pathogens under natural environmental conditions (Ara *et al.*, 2000).

The incorporation of fungicide in the capsule to eradicate contaminants without harming the cultures was an appropriate strategy for the micropropagation (West and Preece, 2006). The inclusion of protectants, bavistin and streptomycin as a constituent of synthetic endosperm had no effect on germination and conversion (Arun Kumar *et al.*, 2005). Ganapathi *et al.* (1992) also found that fungicide and antibiotics have beneficial effects against contamination of the synthetic seeds. In pineapple, the use of antibiotics in the matrix capsule resulted in 86.1% conversion of microshoots (Gangopadyay *et al.*, 2005).

### 6.4.3. Protection from other External Injuries

Naked somatic embryos are very soft and easily susceptible to external injuries under *in vivo* condition. The gel coat of synthetic seed is capable to tolerate the mechanical injuries of machine sowing. The beaded somatic embryos can be used as direct sowing under natural environment.

## 6.5. EASINESS IN TRANSPORTATION

Recent advances in the area synthetic seed technology have revealed that besides somatic embryos, encapsulation of cells and somatic tissues obtained following tissue culture techniques has become popular as a simple way of handling cells and tissues and as an efficient delivery system. Propagating material for many crops is voluminous, for example banana, pineapple, mango, potato etc. Heavy cost involves in transportation and delivery of these seed materials. An interesting strategy for the delivery of these propagules through synthetic seeds is convenient and cost effective. It was noted that for short distance or short-term storage (10 days) at 4 °C there is no reduction in germination of synthetic seeds. The potential uses of artificial seeds in delivery system of plant propagules are:

- Reduced cost of transportation
- Direct greenhouse or field delivery of elite and selected genotypes, hand pollinated hybrid, genetically engineered plants, sterile and unstable genotypes.
- Large-scale mono-culture
- Mixed genotype planting

## 6.6. GERMPLASM CONSERVATION

Besides rapid and mass propagation of plants, the artificial seed technology has added new dimensions not only to handling and transplantations but also for conservation of endangered and precious plant propagules. Currently seed banks are maintained as live plants in the field. This method of conservation is very expensive and dangerous, as it is exposed to natural disasters. The use of artificial seeds would retain these clones in a small space, under controlled conditions (cryopreservation) and without the danger of natural disasters. In addition, this system of germplasm conservation would be particularly useful in tropical species where conservation means are inadequate or nonexistent. The vine (*Vitis* spp.) is a practical example of this system of conservation (Ravi and Anand, 2012).

### 6.6.1. Conservation at Low Temperature

Zygotic embryos are formed from the sexual recombination of male and female gametes. Thus, cuttings or other vegetative means are used for propagation and these methods rarely present convenient storage. Artificial seeds could be a good tool to propagate these types of plants and to store their propagules for a reasonable period of time (Rihan *et al.*, 2011).

Germplasm conservation in clonal crops, particularly root crops, tuber crops and trees associated with many problems. Conservation through tissue culture, after a period of time, it becomes necessary to transfer to fresh media and the sub-culture is a repeated and continuous process. Repeated sub-culture overtime may reduce the morphogenic competence of differentiated cultures (Lynch and Benson, 1991). Therefore, new culture has to be regularly initiated and characterized in order to maintain a regular supply. This approach is highly tedious and costly. The technology of synthetic seed production may help in this regard. Storage of encapsulated embryos for a considerable time allows preservation of valuable, elite gremplasm. Many authors successfully stored synthetic seeds at low temperature (4 °C) for varying periods (Datta *et al.*, 1999; Madhav *et al.*, 2002; Ipekci and Gozukirmizi, 2003; Ahmed and Talukdar, 2005). Pintos *et al.* (2008) also successfully stored synthetic seeds of cork oak at 4 °C for two months without significant loss in conversion capacity. The hydrated synthetic seeds could be stored using low temperature for a few weeks (Redebaugh *et al.*, 1986; Fujii *et al.*, 1989 and 1992). The capability of prolonged storage was achieved when somatic embryos could be dried to moisture content less than 20% (McKersei *et al.*, 1989).

The findings of Taha *et al.* (2012) showed that the encapsulated microshoots with MS medium supplemented with 0.1mg/L BAP + 0.1 mg/L NAA, 3% sucrose, and 0.8% agar gave 93.33% germination without storage. The viability of seeds had fallen from 93.33% to 3.33% after one month storage at 4 °C. Storage conditions such as temperature and period of storage are important factors to determine the regeneration frequency of the stored encapsulated propagules. Storage of synthetic seeds using an alginate encapsulation protocol has been attempted in a few species, with minimal success (Redenbaug *et al.* 1986, 1987; Fuji *et al.*, 1987). The synthetic seeds of sweet corn were germinated to 43 and 55% after 2 weeks of storage under 15 ±2°C and 25 ±2 °C, respectively (Thobunluepop, 2009). In the present study, low germination rate (16.67%) was recorded after 2 weeks of storage at 4 °C. Therefore, further research for the development of a better technique is required. For example, increasing the storage temperature may increase regeneration levels. Elvax 4260 (ethylene vinyl acetate acrylic acid terpolymer, Du Pont, USA) could be used for coating the capsules to avoid rapid water loss when calcium alginate capsules are exposed to the ambient atmosphere (Ara *et al.*, 2000). Generally, 4 °C has been found to be most suitable for artificial seeds storage (Saiprasad and Polisetty, 2003; Ikhlaq *et al.*, 2010) although there is a wealth of literature on the choice of optimum temperature and light conditions.

Beside rapid and mass propagation of plants, artificial seed technology added new dimensions in conservation of endangered and desirable genotypes. Datta *et al.*, (1999) developed protocol for the endangered orchid *Geodorum densiflorum* to *in vivo* germination and regeneration of plantlet for storage and transplantation.

### 6.6.2. Conservation in Tissue Culture Room

The alginate coated beads, made by encapsulating small propagules are excellent stable germplasm units. Bhattacharyya *et al.* (2007) successfully stored synthetic seed of *Plumbago indica*, a medicinal plant at culture room conditions (22-24 °C temperature and in dark) for 3 months without significance reduction of conversion ability.

### 6.6.3. Cryopreservation

***Cryopreservation*** is commonly used technique for long-term preservation of biological material. This can be defined to the stepwise viable freezing of biological materials (seed, planting materials, plant callus, somatic embryos, synthetic seeds etc.) followed by storage at ultra-low temperature, preferably at that of liquid nitrogen (–196 °C). In Greece, *Kryos* means frost, thus cryopreservation is the preservation in the frozen state. This process preserves growth and biosynthetic potencies of biological materials. It arrests all metabolic activities and biological deterioration of cells, thus the material can be preserved for longer period of time. Practically, it can be stored on solid $CO_2$ (–79 °C), in deep freezers (–80 °C or above), in vapour phase of nitrogen (–150 °C) or in liquid nitrogen (–196 °C). Cryopreservation is considered as an ideal means of avoiding loss of embryopgenic potential during repeated subcultures and as a means of preventing the occurrence of somaclonal variation during long-term maintenance of embryogenic culture. Cryopreservation of biological material depends mainly on two parameters:

(i) Freezing tolerance ability of the biological material: It is to be remembered that the viability of cells is not affected at the ultra-low temperature while utmost protection is required during freezing and thawing.

(ii) The formation of ice crystal inside the cell (cytoplasm)

Cryopreservation involves a series of steps, namely:

(i) Specimen preparation (synthetic seeds)

(ii) Specimen treatment with ***cryoprotectants***

(iii) Freezing of the biological material (synthetic seeds)

(iv) Storage at ultra-low temperature

The cryopreserved bio-materials must be brought back to it natural state whenever it is desired. Following steps are followed to restore the natural state of the bio-material after cryopreservation:

(i) Takeout the preserved bio-materials (synthetic seeds) from cryopreservation.

(ii) Thawing

(iii) Plant growth and regeneration

### 6.6.4. Encapsulation-vitrification

Encapsulation-vitrification is a new technique of preservation of plant materials, which combines the advantages of vitrification (rapidity of implementation) and of emasculation-dehydration (ease of manipulation of encapsulated explants) has been established (Matsumoto *et al.*, 1995a). This method is user-friendly and greatly reduce the time requires for dehydration. Thus, the method is frequently being used for long term preservation of plant materials (Table 6.1).

**Table 6.1.** Cryopreserved plant species using encapsulation-vitrification technique

| Plant species | Explant* | Reference |
|---|---|---|
| *Ananascomosus* | Shoot tips | Gamez-Pastrana *et al.* (2004) |
| *Armoraciarusticana* | Hairy roots | Phunehindawan *et al.* (1997) |
| *Citrus aurantium* | Shoot tips | Al-Ababneh *et al.* (2002) |
| *Daucuscarota* | Shoot tips | Hirai (2001) |
| *Daucuscarota* | Somatic embryos | Tannoury (1993) |
| *Dianthus caryophyllus* | Shoot tips | Tannoury *et al.* (1991); Tannoury (1993) |
| *Dioscorea* | Shoot tips | Hirai (2001), Hirai and Sakai (2001) |
| *Diospyros kaki* | Shoot tips | Wang *et al.* (1998) |
| *Fragaria×Ananassa* | Shoot tips | Hirai (2001), Hirai and Sakai (2001) |
| *Gentiana* spp. | Shoot tips | Tanaka *et al.* (2004) |
| *Ipomoea batatas* | Shoot tips | Hirai and Sakai (2003) |
| *Lilium* | Shoot tips | Hirai (2001) |
| *Malusdomestica* | Shoot tips | Paul *et al.* (2000) |
| *Monihotesculenta* | Shoot tips | Hirai (2001); Charoensub *et al.* (2004) |
| *Menthaspicata* | Shoot tips | Hirai and Sakai (1999a); Hirai (2001); Hirai and Sakai (2000) |
| *Oleaeuropea* | Somatic embryos | Shibli and Al-Juboory (2000) |
| *Poncirus trifoliate* × *Citrus sinensis* | Shoot tips | Wang *et al.* (2002) |
| *Prunusdomestica* | Shoot tips | DeCarlo *et al.* (2000) |
| *Rubusidaeus* | Shoot tips | Wang *et al.* (2005) |
| *Saintpauliainantha* | Shoot tips | Moges *et al.* (2004) |
| *Solanumtuberosum* | Shoot tips | Hirai and Sakai (1999b) Hirai (2001); Hirai and Sakai (2001) |
| *Vitis*spp. | Shoot tips | Wang *et al.* (2004) |
| *Vitisberlandieri* × *V. riparia* | Shoot tips | Benelli *et al.* (2003) |
| *Wasabia japonica* | Shoot tips | Matsumoto *et al.* (1995b) |

*Source*: Sakai and Engelmann (2007)

*Explant used for synthetic seed preparation

### 6.6.5. Encapsulation-dehydration

Encapsulation-dehydration is a valuable procedure for various plant materials, including *in vitro*-grown shoot tips and somatic embryos. In this method, the synthetic seeds are treated with a high sucrose concentration, dried down to moisture content 20-30% (under airflow or using silica gel) and subsequently frozen in liquid nitrogen (Fabre and Dereuddre, 1990). This technique may prove interesting in two situations- (1) for the materials which are recalcitrant to standard freezing techniques, (2) the seeds that the protection conferred by the beads allows submitting the embedded material to pretreatment conditions which would otherwise be detrimental. It may be beneficial also if the encapsulation allows to carryout rapid freezing, thus avoiding the use of a programmable freezing apparatus and simplifying the process. The drawback of this procedure is that it is rather lengthy and labour-intensive.

## References


Ahmed H, Talukdar MC. 2005. *In vitro* propagation and artificial seed production of *Arundinabambusiflolia*. J Ornamental Hort New-Series. 8(4): 281-283.

Al-Ababneh SS, Karam NS, Shibli RA. 2002. Cryopreservation of sour orange (*Citrus auratium* L.) shoot tips.*In Vitro* Cell Develop Biol- Plant. 38: 602-607.

Ara H, Jaiswal U, Jaiswal VS. Synthetic seed: prospects and limitations. 2000. Current Sci. 78(12): 1438- 1444.

Arun Kumar MB, Vakeswaran V, Krishnasamy V. 2005. Enhancement of synthetic seed conversion to seedlings in hybrid rice. Plant Cell Tiss Org. 81: 97-100.

Attree SM, Fowke LC. 1993. Embryogeny of gymnosperms: Advances in synthetic seed technology of conifers. Plant Cell Tiss. Org. Cult. 35 (1): 1-35.

Bekheet SA. 2006. A synthetic seed method through encapsulation of *in vitro* proliferated bulblets of garlic (*Allium sativum* L.). Arab J Biotech. 9(3): 415-426.

Benelli C, Lambardi M, Fabbri A. 2003. Low of axillary shoot tips of In vitro grown grape (Vitis) temperature storage and cryopreservation of the by a two stepvitrification protocol. Euphytica, 131: grape rootstock “Kober 5BB”. ActaHorticulturae. 623: 249-253.

Bhattacharyya R, Ray A, Gangopadhyay M, Bhattachaya S. 2007. *In vitro* conservation of *Plumbagoindica*- a rare medicinal plant. Plant Cell BiotechnolMol Biol. 8(1/2): 39-46.

Bordallo PN, Silva DH, Maria J, Cruz CD, Fontes EP. 2004. Somaclonal variation on *in vitro* callus culture potato cultivars. Horticultura-Brasileira. 22(2): 300-304.

Brar DS, Fujimura T, McCouch S, Zapate FJ. 1994. Application of biotechnology in hybrid rice. In: Hybrid Rice Technology: New Development and Future Prospects, Virmani SS (ed.), International Rice Research Institute, Manila, Philippines. pp. 51-62.

Burschel P, Gallegos C, Martinez O, Holl YW. 1976. Composición y dinámica regenerative de un bosque virgin mixto de rauli y coigüe. Universidad Austral de Chile, Valdivia. Bosques (Chile). 1: 55-74.

Cangahuala-Inocente GC, Vesco LL-dal, Steminmacher D, Torrres AC, Guerra MP. 2007. Improvement in somatic embryogenesis protocol in Feijo (*Accasellowian* (Berg) Burret): introduction, conversion and synthetic seeds.

Cartes P, Castellanos H, Ríos D, Sáez K, Spierccolli S, Sánchez M. 2009. Encapsulated somatic embryos and zygotic embryos for obtaining artificial seeds of rauli-beech [*Nothofagusalpina* (Poepp. & Endl.) oerst.]. Chil J Agric Res. 69: 112-118.

Cartes PR, Castellanos HB, Rios DL, Sáez KC, Spierccolli SH, Sánchez MO. 2009. Encapsulated somatic and zygotic embryos for for obtaining artificial seeds of rauli-beech (*Nothofagus alpine* (Poepp. & Endli.) Oerst.). Chilean J Agri Res. 69(1): 112-118.

Charoensub R, Hirai D, Sakai A. 2004. Cryopreservation of *in vitro*-grown shoot tips of cassava by encapsulation-vitrification method. Cryo-Letters. 25: 51-58.

Danso KE, Ford-Flyood BV. 2003. Encapsulation of nodal cuttings and shoot tips for storage and exchange of Cassava germplasm. Plant Cell Rep. 21: 718-725.

Datta KB, Kanjilal B, Sarker D-de, De-Sarkar D. 1999. Artificial seed technology: Development of a protocol in *Geodorumdensiflorum* (Lam) Schltr. – an endangered orchid. Current Sci. 76(8): 1142-1145.

Daud M, Taha MZ, Hasbullah AZ. 2008. Artificial seed production from encapsulated micro shoots of *Sainpauliaionantha*Wendl. (African Violet). J Appl Sci. 8: 4662-4667.

DeCarlo A, Benelli C, Lambardi M. 2000. Development of a shoot-tip vitrification protocol and comparison with encapsulation-based procedures for plum (*Prunusdomestica*L.) cryopreservation. Cryo-Letters. 21: 215-222.

Desai BB, Kotecha PM, Salukhe DK. 1997. Seeds Handbook- Biology, Production, Processing and Storage. pp. 91-113.

El-Halwgy AA. Youssef SS, Abou-Zaied A, Hussein MH. 2004. Molecular and chromosomal stability of cryopreserved almond tissues. ProcIntConf Genet Eng. Appl. (April 8-11, 2004). pp. 279-298.

El-Kazzaz AA, Taha HS. 2002. Tissue culture of broccoli and molecular characterization. Bull NRC Egypt. 27(4): 481-490.

Fabre J, Dereuddre J. 1990. Encapsulation-dehydration: A new approach to conservation of *Solanum* shoot-tips. Cryo-Letters. 11: 413-426.

Fujii JA, Slade D, Aguirre Rascon J, Redebaugh K. 1992. Field Planning alfalfa artificial seeds. *In vitro* Cell Dev Biol. 18P: 73-80.

Fujii JA, Slade D. Redenbaugh K. 1989. Maturation and greenhouse planting alfalfa artificial seeds. *In vitro* Cell Dev. Biol. 25:1179-1182.

Gamez-Pastrana R, Martinez-Ocampo Y, Beristain CI, Gonzalez-Amao MT. 2004. An improved cryopreservation protocol for pineapple apices using encapsulation-vitrification. Cryo-Letters. 25: 405-414.

Ganapathi TR, Suprasanna P, Bapat VA, Rao PS. 1992. Propagation of banana through encapsulated shoot tips. Plant Cell Rep. 11: 571-575.

Gangopadyay G, Bandopadyaya T, Podder R, Gangopadhyaya SB, Mukherjee KK. 2005. Encapsulation of pineapple microshoots in alginate beads for temporary storage. Current Sci. 88: 972-977.

Gray DJ, Purohit A. 1991. Somatic embryogenesis and development of synthetic seed technology. Critical Reviews Plant Sci. 10(1): 33-61.

Gray DJ. 1997. Synthetic for Clonal Seed Production of Crop Plants. In: R.B. Taylorson (ed.) Recent Advances in the Development and Germination of Seeds. pp. 29-45.

Hirai D, Sakai A. 1999a. Cryopreservation of *in vitro*-grown axilary shoot-tip meristems of mint (*Menthaspicata* L.) by encapsulation vitrification. Plant Cell Reports. 19: 150-155.

Hirai D, Sakai A. 1999b. Cryopreservation of *in vitro*-grown meristem of potato (*Solanumtuberosum* L.) by encapsulation-vitrification.Potato Research. 42: 153-160.

Hirai D, Sakai A. 2001. Recovery growth of plants cryopreserved by encapsulation – vitrification. Bull Hokkaido Prefectural Agricultural Experiment Station. 80: 55-64.

Hirai D, Sakai A. 2003. Simplified cryopreservation of sweet potato [*Ipomoea batatas* (L.) Lam] by optimizing conditions for osmoprotection.Plant Cell Reports. 21: 961-966.

Hirai D. 2001. Studies on cryopreservation of vegetatively propagated crops by encapsulation vitrification method. In: Rep Hokkaido Prefectural Agricultural Experiment Station. 99: 58.

Ikhlaq M, Hafiz IA, Micheli M, Ahmad T, Abbasi NA, Standardi A. 2010. *In vitro* storage of synthetic seeds: effect of different storage conditions and intervals on their conversion ability. African J Biotech. 9: 5712-5721.

Ipekci Z, Gozukirmizi N. 2003. Direct somatic embryogenesis and synthetic seed production from *Paulownia elongate*. Plant Cell Rep. 22(1): 16-24.

King MW, Roberts. 1980. Maintenance of recalcitrant seeds in storage. In: Recalcitrant crop seeds. Chin, H.F., Robert, E.H. (eds.), Tropical Press Sdn. Bhd. 29, JalanRiong, Kuala Lumpur, 22-03 Malaysia.

Kumar U. 1998. Synthetic Seeds for Commercial Crop Production. 160 pp. Vedams Books (P) Ltd. New Delhi, India.

Lynch PT, Benson EE. 1991. Cryopreservation, a method for maintaining the plant regeneration capacity of rice, cell suspension culture. In: Proceedings of the Second Rice Genetic Symposium, International Rice Research Institute, Los Banos, Philippines. pp. 321-332.

Madhav MS, Rao BN, Singh S, Deba PC. 2002. Nucellar embryogenesis and artificial seed production in *Citrus reticulata*. Plant Cell BiotechnolMol Biol. 3(1-2): 77-80.

Matsumoto K, Matsumoto K, Hirao C, Teixeira J. 1995a. *In vitro* growth of encapsulated shoot tips in banana (*Musa* sp.). ActaHorticulturae. 370: 13-20.

Matsumoto T, Sakai A, Takahashi C, Yamada K. 1995b. Cryopreservation of in vitro-grown meristems of wasabi (Wasabia japonica) by encapsulation-vitrification method. Cryo-Letters. 16: 189-196.

McKersie BD, Bowley SR. 1993. Synthetic seeds in alfalfa. In: Synseeds Application of Synthetic Seeds to Crop Improvement, Redenbaugh K (ed.). CRC Press, Boca Raton. pp. 231-255.

McKersie BD, Brown DCW. 1996. Somatic embryogenesis and artificial seeds in forage legumes. Seed Sci Res.6: 109-126.

McKersie BD, Brown DCW. 1997. Biotechnology and the Improvement of Forage Legumes. CAB International, Wallingford, Oxon, UK., pp. 111-132.

McKersie BD, Senaratna T, Bowley SR, Brown DCW, Krochko JE, Bewley JD. 1989. Application of artificial seed technology in the production of hybrid alfalfa (*Medicago sativa* L.). In Vitro Cell Dev Biol. 25: 1183-1188.

McKersie BD, Senaratna T, Bowley SR, Brown DCW, Krochko JE, Bewley JD. 1989. Application of artificial seed technology in the production of hybrid alfalfa (*Medicago sativa* L.). In Vitro Cell Dev Biol. 25: 1183-1188.

Moges AD, Shibli RA, Karam NS. 2004. Cryopreservation of African violet (*Saintpaulaionantha* Wendl.) shoot tips. *In Vitro* Cell Develop Biotechnoll- Plant. 40: 389-395.

Murashige T. 1978. The impact of plant tissue culture on agriculture. In: Frontiers in Plant Tissue Culture, Thrope TA (ed.). Calgary University Press, Calgary, Alberta, Canada.

Naik SK, Chand PK. 2006. Nutrient-alginate encapsulation of *in vitro* nodal segments of pomegranate (*Punicagranatum* L.) for germplasm distribution and exchange. ScientiaHorticulturae. 108: 247-252.

Nower AA, Ali EAM, Rizakalla AA. 2007. Synthetic seeds of pear (*Pyruscommunis* L.) rootstocks storage *in vitro*. Australian J Basic Appl Sci. 1(3): 262-270.

Nyende AB, Schittenhelm S, Mix-Wagner G, Greef JM. 2005. Yield and canopy growth characteristics of field grown synthetic seeds of potato. European J Agronomy. 22:175-184.

Nyende AB, Schittenhelm S, Wagner GM, Greef JM. 2003. Production, storability, and regeneration of shoot tips of potato (*Solanumtuberosum* L.) encapsulated in calcium alginate hollow beads. In Vitro Cell DevBiol Plant. 39:540-544.

Onishi N, Sakamoto Y, Hirosawa T. 1994. Synthetic seeds as an application of mass production of embryos. Plant Cell Tiss Org Cult.39: 137-145.

Park YS, Barrett JD, Bonga JM. 1998. *In Vitro* Cellular Devl. Biol Plant. 34(3):231-239.

Paul H, Daigny G, Sangwan-Noreel BS. 2000. Cryopreservation of apple (*Malus×domestica* Borkh.) shoot tips following emasculation-dehydration or emusculation-vitrification. Plant Cell Reports. 19: 768-774.

Pellinen TP, SorvarisTahronen R, Sewon P. 1997. Somatic embryogenesis in cucumber (*Cucumissativus* L.) callus and suspension cultures. Angowandte Botanik. 71(3-4): 116-118.

Phunchindawan M, Hirata K, Miyamoto K, Sakai A. 1997. Cryopreservation of encapsulated shoot primodia induced in horseradish (*Armoraciarusticana*) hairy root cultures. Plant Cell Reports. 16: 469-473.

Pintos B, Bueno MA, Cuenca B, Manzanera JA. 2008. Synthetic seed production from encapsulated somatic embryos of cork oak (*Quercussuber* L.) and automated growth monitoring. Plant Cell Tiss Organ Cult. 95(2): 217-225.

Rady MR, Hanafy MS. 2004. Synthetic seed technology for emasculation and regrowth of *in vitro*-derived *Gypsophila paniculata* L. shoot-tips. Arab J Biotech. 7(2): 251-264.

Ravi D, Anand P. 2012. Production and Applications of Artificial seeds: A Review. Inter Res J Biol Sci. 1(5): 74-78.

Redenbaugh K, 1993. Synseeds: Application of synthetic seeds to crop improvement. CRC Press, Boca Raton.

Redenbaugh K, Fujii JA, Slade D. 1991. Synthetic seed technology. In: Cell Culture and Somatic Cell Geneitcs in Plants, Vol. 8, Vasil, I.K. (ed.), Academic Press, New York. pp. 35-74.

Redenbaugh K, Paasch BD, Nichol JW, Kossler ME, Viss PR, Walkee KA. 1986. Somatic seed: encapsulation of asexual plant embryos. Biotechnology. 4: 797-801.

Redenbaugh K, Slade D, Viss P, Fujii JA. 1987. Encapsulation of somatic embryos in synthetic seed coats. Hortscience. 22: 803-809.

Rihan HZ, Al-Issawi M, Burchett S, Fuller MP. 2011. Encapsulation of cauliflower (*Brassica oleracea*var botrytis) microshoots as artificial seeds and their conversion and growth in commercial substrates. Plant Cell Tiss Organ Cult (2011) 107:243-250.

Saiprasad G. 2001. Artificial seeds and their applications. Resonance. 6: 39-47.

Saiprasad GVS, Polisetty R. 2003. Propagation of three orchid genera using encapsulated protocorm-like bodies. *In vitro* Cell DevBiol–Plant. 3: 42-48.

Sakai A. 2000. Development of cryopreservation techniques. In: Engelmann F, Takagi H (eds) Cryopreservation of tropical plant germplasm: current research progress and application. JIRCAS, Tsukuba/IPGRI, Rome. pp 1-7.

Sakamoto Y, Ohnishi N, Hayashi M, Okamoto A, Mashiko T, Sanada M. 1991. Synthetic seeds: the development of a botanical seed analogue. Chemical Regulation of Plants. 26(2): 205-211.

Shibli RA, Al-Juboory KH. 2000. Cryopreservation of 'Nabali' olive (*Oleaeuropea* L.) somatic embryos by encapsulation-dehydration and encapsulation-dehydration. Cryo-Letters. 21: 357-366.

Singh AK. 2006. Flower crops: Cultivation and management. New India Publishing, New Delhi, India, ISBN: 8189422359. pp. 429.

Sudhkar K, Nagraj BN, Santhoshkumar AV, Sunilkumar KK, Vijayakumar NK. 2000. Studies on the production and storage potential of synthetic seeds in cocoa (*Theobroma cocoa* L.). Seed Res. 28(2): 119-125.

Sunilkumar KK, Sudhakar K, Vijayakumar NK. 2000. An attempt to improve storage life of *Hopeaparviflora* seeds through synthetic seed production. Seed Res. 28(2) : 126-130.

Taha RM, Saleh A, Mahmad N, Hasbullah NA, Mohajer S. 2012. Germination and Plantlet Regeneration of Encapsulated Microshoots of Aromatic Rice (*Oryza sativa* L. Cv.MRQ 74). The Scientific World J. 2: 1-6.

Tanaka D, Niino T,Isuzugawa K, Hikage T, Uemura M. 2004. Cryopreservation of shoot apices of *in vitro*-grown gentian plants: comparison of vitrification and encapsulation-vitrification protocol.Cryo-Letters. 25: 167-176.

Tannoury M, Ralambosoa J, Kaminski M, Dereuddre J. 1991. ComptesRendus de I'Académie des Sciences Paris. 313: Série III. pp. 633-638.

Tannoury M. 1993. Cryoconservationd'Apexd'Oeillet (*Dianthus caryophyllus* L.) et d'Embryons Somatiques de Carotte (*Dacuscarota* L.) par les Procédésd'Enrobage-Déshydratation et d'Enrobage-Vitrification. Thésed'Université, Université Paris 6.

Thobunluepop P, Pawelzik E, Vearasilp S. 2009. Possibility of sweet corn synthetic seed production. Pakistan J Biol Sci. 12(15): 1085-1089.

Tian I, Brown DC. 2000. Improvement of soybean somatic embryo Development and Maturation by abscisic acid steam treatment. Can J Plant Sci. 80: 721-726.

Towill LE. 1988. Genetic considerations for clonal germplasm preservation of materials. HortScience. 23: 91-93.

Wang JH, Liu F, Huang CN, Yan QS, Zhang XQ. 1998. Chinese Rice Research Newsletter. 6: 5-6.

Wang QC, Batuman Ö, Li P, Bar-Joseph M, Gafny R. 2002. A simplified efficient cryopreservation of *in vitro* grown shoot tips of 'Troyer" citrange [*Poncirus trifoliate* (L.) Raf. ×*Citrus sinensis* (L.) Osbeck] by encapsulation-vitrification. Euphytica. 128: 135-142.

Wang QC, Laamanen J, Uosukainen M, Volkonen JPT. 2005. Cryopreservation of *in vitro*-grown shoot tips of raspberry (*Rubusidaeus* L.) by encapsulation-vitrification and emasculation dehydration. Plant Cell Reports. 24: 280-288.

Wang QC, Mawassi M, Sahar N, Li P, Violeta CT, Gafny R, Sela I, Tanne E, Perl A. 2004. Cryopreservation of Grapevine (*Vitis* spp.) Embryogenic Cell Suspensions by Encapsulation–Vitrification. Plant Cell Tiss Org Cult. 77: 267-275.

West TP, Preece JE. 2006. Use of Acephate, Benomy and alginate encapsulation for eliminating culture mites and fungal contamination from *in vitro* culture of hardy hibiscus (*Hibiscus moscheutos* L.). *In vitro* Cell DevBiol- Plant. 42: 301-304.

Winkelmann T, Meyer L, Serek M. 2004. Germination of encapsulated somatic embryos of *Cyclamen persicum*. HortScience.39: 1093-1097.

Ye KN, Huang JC, Ji BJ, You CB, Chen ZL, Ding Y. 1993. Hybrid papaya artificial seed production for experimental field. Current Plant SciBiotechnol Agri. 15: 411-413.

# 7

# Hardening of Artificial Seed Derived Plantlets

Plant tissue culture is mainly referred to as mass-multiplication of *in vitro* derived plantlets grown under controlled environment. Subsequently, the production of artificial seeds from *in vitro* grown plant propagules is also a method of mass-production of true-to-type plantlets. Micro-propagation allows rapid and mass production of high quality, disease free and uniform planting material irrespective of the season and weather. Micro-shoots on being transferred to *ex vitro* conditions are exposed to abiotic (altered temperature, light intensity and humidity conditions) and biotic stress conditions i.e. soil micro-flora, so there is need for acclimatization for successful establishment and survival of plantlets (Deb and Imchen, 2010). In order to increase growth and reduce mortality in plantlets at the acclimatisation stage, research has been focused on the control of the environmental conditions (both physical and chemical) and to acclimatize the plants to compete with soil micro-flora (Mathur *et al*., 2008). In this chapter methods of acclimatization of plantlets derived from simple plant tissue culture or from synthetic seeds are discussed.

## 7.1. NEED OF ACCLIMATIZATION TO *EX VITRO* CONDITIONS

*Ex vitro* (Latin: 'out of the living') means that which takes place outside an organism or under controlled environment. *Ex vitro* conditions allow experimentation on an organism's cells or tissues under more controlled conditionsthen only it is possible in *in vitro* experiments (in the intact organism or controlled environment), at the expense of altering the 'natural' environment. This alteration of natural environment alters some of the morphology and/or physiology of *in vitro* developed plantlets.

Some of the alterations that occur duringin vitrogrown plantlets are being described in this chapter.

### 7.1.1. Induced Altered Morphology

*In vitro* grown plantlets are different from naturally grown plants under *in vivo* condition. Sufficient mortality is observed upon transfer of micro-shoots to *ex vitro* conditions as the cultured plants have non-functional stomata, weak root system and poorly developed cuticle (Mathur *et al.*, 2008). Plantlets supplied with an excess of phytohormones shows abnormalities in ***morphology*** and anatomy and are called vitrified plants plants (Hronkova *et al.*, 2003). The physiological and anatomical characteristics of micro-propagated plantlets necessitate that they should be gradually acclimatized to field condition (Hazarika, 2003).

#### *7.1.1.1. Leaf Cuticle*

In many plant species, the leaves formed under*in vitro* are unable to develop further under *ex vitro* conditions and they are replaced by newly formed leaves (Preece and Sutter, 1991; Diettrich *et al.*, 1992). Leaves from *in vitro* plantlets of *Liquidambar styraciflua* had a less developed cuticle in contrary to the well-developed cuticle in leaves of transplanted and field-grown plants (Wetzstein and Sommer, 1982). An increase in cuticle thickness, mass and wax content from young to adult leaves was found in both *in vitro* and *ex vitro* grown Hedera helix plants (Gilly *et al.*, 1997). Development of cuticle, epicuticular waxes, and effective stomatal regulation of transpiration occurs leading to stabilization of water potential of field transferred plantlets (Pospisilova *et al.*, 1999). The structure and quantity of epicuticular waxes on the upper surface of *Brassica oleracea* leaves, determined two weeks after transplantation, were similar to those on seedling leaves (Grout and Aston, 1977). The relative wax content on an area basis of *Liquidambar styraciflua* leaves remained unchanged whereas that in *Malus domestica* leaves decreased after acclimatization (Sutter, 1988). On the other hand, in *Malus pumila* plantlets, thickness of epicuticular waxes of leaves formed under*in vitro* were not affected after being transfereed to *ex vitro* conditions, but it was higher in newly formed leaves (Díaz-Pérez *et al.*, 1995b).

#### *7.1.1.2. Stomata*

In *Liquidambar styraciflua*, *Vacciniumcorymbosum* and *Nicotiana tabacum* stomatal density decreased after transplantation (Wetzstein and Sommer 1983, Noé and Bonini 1996, Tichá et al. 1999). *In vitro* grown plantlets have large stomata with changed shape and structure. Guard cells have thinner cell walls and contain more starch and chloroplast (Marin *et al.*, 1988). Ticha *et al.* (1999) found that acclimatization of tobacco plantlets to *ex vitro* conditions decreased stomata density and also changed the size and morphology of stomata on both sides of newly formed leaves. During acclimatization to *ex vitro* conditions, leaf thickness generally increases, leaf mesophyll progresses in differentiation into palisade and spongy parenchyma, stomatal density decreases and stomatal form changes from circular to elliptical one.

#### *7.1.1.3. Epidermal Hairs*

In *Rubus idaeus*, the number of epidermal hairs was low *in vitro*, higher in leaves formed after transplantation, and the highest in greenhouse and field grown plants (Donnelly and Vidaver, 1984).in *Prunus serotina* and *Rhododendron* spp. plants stomatal density increased and stomata pore length decreased after transplantation (Waldenmaier and Schmidt, 1990, Drew *et al.*, 1992). Leaves from *in vitro* grown *Prunus cerasus, Vaccinium corymbosum* or *Nicotiana tabacum* plantlets had ring-shaped stomata, but in leaves of *ex vitro* transferred plants stomata were elliptical (Marín *et al.*, 1988, Noé and Bonini, 1996, Tichá *et al.*, 1999).

#### *7.1.1.4. Vacuoles*

Mesophyll cells of *in vitro* grown *Liquidambar styraciflua* plantlets had large vacuoles, limited cytoplasmic content, and flattened chloroplasts with irregularly arranged internal membrane systems (Wetzstein and Sommer, 1982); internal chloroplast organisation was strongly light dependent (Lee *et al.*, 1985). On the contrary, the chloroplasts of acclimatized *Liquidambar styraciflua* plants had well developed grana, osmiophillic globules and frequently starch granules (Wetzstein and Sommer, 1982).

#### *7.1.1.5. Leaf Tissue*

In *Nicotiana tabacum* plantlets grown under *in vitro*, the leaf mesophyll consisted of one layer of loosened, poorly differentiated palisade parenchyma and two to three layers of spongy parenchyma (Tichá and Kutík, 1992), and in *Vaccinium corymbosum* of only one or two layers of disorganised spongy parenchyma (Noé and Bonini, 1996). Large intercellular spaces were present in the mesophyll (Brainerd *et al.*, 1981; Dami and Hughes, 1995; Noé and Bonini, 1996). In *Brassica oleracea*, one layer of palisade mesophyll cells was developed three weeks after transplantation into a greenhouse (Grout and Aston, 1978).

Leaves are thicker in acclimatized *Liquidambar styraciflua* plants than in *in vitro* grown plantlets, and mesophyll tissue differentiated into palisade and spongy parenchyma were found; the spongy parenchyma had fewer and smaller air-spaces (Wetzstein and Sommer, 1982). Similar results were found by Brainerd *et al.* (1981) in *Prunusinsititia*, Donnelly and Vidaver (1984) in *Rubusidaeus*, Lee *et al.* (1988) in *Liquidambar styraciflua*, Waldenmaier and Schmidt (1990) in *Rhododendron* spp., Johansson *et al.* (1992) in *Rosa odorata* × *Rosa damascena*, and Noé and Bonini (1996) in *Vaccinium corymbosum*.

*7.1.1.6. Leaf Cell Organelles*

Guard cells of *in vitro* grown *Rosa hybrida* plantlets contained numerous ribosomes and mitochondria, starch-rich plastids, and relatively large vacuoles indicating that they may exhibit metabolic activity similar to normal guard cells. In guard cells of *in vitro* grown *Solanum phureja* plantlets, consistently more chloroplasts were found than in those of *ex vitro* grown plants (Singsit and Veilleux, 1991). In *Prunuscerasifera* plantlets, the ability of stomata to close in response to abiotic factors and to re-open after treatment was much higher in young than in adult and old leaves which might be also important during transfer to *ex vitro* conditions (Zacchini and Morini, 1998).

### 7.1.2. Induced Altered Physiology

*7.1.2.1. Chlorophyll Contents*

Chlorophyll-*a* and Chlorophyll-*b* contents increased after transplantation (Synková 1997;Pospíšilováet *al.*, 1998). The same effect was evident in originally ***photoautotrophically*** grown *Nicotiana tabacum* plants but in originally ***photomixotrophically*** grown plants, an abrupt decrease in Chl-*a* and Chl-*b* contents during the first week after transplantation followed by a slow increase was found (Kadlecek, 1997). Chl-*a* fluorescence parameters (e.g., variable to maximum fluorescence ratio $F_v$/F, actual quantum yield of photosystem-II) increased after transfer of *Elaeis guineensis* and *Nicotiana tabacum* plantlets to *ex vitro* conditions (Synková, 1997).

*7.1.2.2. Photosynthetic Rate*

Net photosynthetic rate ($P_N$) in *Solanum tuberosum* and *Spathiphyllum floribundum* plants decreased in the first week after transplantation and increased thereafter (Baroja, 1993; Baroja *et al.*,1995, Van Huylenbroeck and Debergh, 1996). After transplantation, the $^{14}CO_2$ uptake by persistent leaves of *Fragaria* and *Rubus idaeus* was similar to that in plantlets grown *in vitro* or was slightly increased, and a significantly increased $^{14}CO_2$ uptake was found only in newly formed leaves (Short *et al.*, 1984, Deng and Donnelly, 1993). In *Calathea louisae* the *in vitro* formed leaves were not able to photosynthesize during the first days after transfer, but in *Spathiphyllum floribundum*, the *in vitro* formed leaves were photosynthetically competent and normal source-sink relations were observed. Nevertheless, in both plant species, substantial photosynthetic activities were measured when new leaves were fully developed (Van Huylenbroeck *et al.*, 1998). Two weeks after *ex vitro*transplantation of *Nicotiana tabacum* plantlets, $P_N$, maximum photochemical efficiency, and actual quantum yield of photosystem-II were higher than in plantlets grown *in vitro* (Pospíšilová *et al.*, 1998). Similarly, higher $P_N$ was found in *Maluspumila* plants three weeks after transplantation (Díaz-Pérez *et al.*, 1995b) and more than twice as high a maximum $P_N$ in *Vitisvinifera* × *Vitisberlandieri* rootstocks one month after transplantation (Fila *et al.*, 1998). Exposure of *Calathealouisae* and *Spathiphyllum floribundum* plantlets to high irradiance immediately after transplantation caused photoinhibition and even Chlphotobleaching (Van Huylenbroeck, 1994, Van Huylenbroeck *et al.*, 1995). The above mentioned results suggest that photoinhibition might be the cause of the transient decrease in photosynthesis after transplantation. However, no photoinhibition was observed in plants acclimatized under low irradiance for four weeks (Van Huylenbroeck, 1994). Similarly, photoinhibition was observed in *Rosa hybrida* plantlets, but only in the first week after *ex vitro* transfer, and especially in those plantlets transplanted into medium with decreased osmotic potential by addition of mannitol (Sallanon *et al.*, 1998). In *Nicotiana tabacum*, plantlets were acclimatized to *ex vitro* conditions under slight shade in a greenhouse where irradiance varied during the day and daily maximum was usually less than that needed for saturation of photosynthesis, no photoinhibition occurred: $F_v/F_m$ was in the range typical for non-stressed plants and did not change during acclimatization. The degree of de-epoxidation of xanthophyll cycle pigments was not changed (Pospíšilová *et al.*, 1999).

### 7.1.3. Acclimatization to Abiotic Stresses

#### *7.1.3.1. Light Intensity*

*In vitro* growing plantlets are under low light intensity (1200-3000 lux) and temperature (25 ± 2°C), hence direct transfer to broad spectrum sunlight (4000-12000 lux) and temperature (26-36°C) might cause charring of leaves and wilting of plantlets. It is therefore necessary to accustom the plant in the natural conditions by a process of hardening or acclimatization (Lavanya *et al.*, 2009). The culture containers could be kept in the greenhouse with loose lids. Micro propagated plantlets can be left in shade for 3-6 days under diffused natural light to make them adjust to the conditions of new environment. This helps in semi-hardening of plants and leads to shoot elongation. Transfer of microshoots from *in vitro* to *ex vitro* conditions under direct sunlight might cause photoinhibition and chlorophyll (Chl) photobleaching. Study shows that exposure of *Calathealouisae* and *Spathiphyllumfloribundum* plantlets to high irradiance immediately after transplantation caused photoinhibition and even Chlphotobleaching (Van Huylenbroeck, 1994; Van Huylenbroeck *et al.*, 1995). No photoinhibition was found during growth in the greenhouse when *Nicotianatabacum* plantlets were acclimatized in two phases, first in the greenhouse (low irradiance of 30-90 lmol $m^{-2}$ $s^{-1}$) and then in the open air (200-1400 lmol $m^{-2}$ $s^{-1}$) (Posplsilova *et al.*, 1999). *In vitro* to *ex vitro* transfer might lead to a transient decrease in photosynthetic parameters. Net photosynthetic rate in *Solanumtuberosum* and *Spathiphyllumfloribundum* plants decreased in the 1st week after transplantation and increased thereafter (Pospisilova *et al.*, 1999). After transplantation, the $^{14}CO_2$ uptake by persistent leaves of *Fragaria* and *Rubusidaeus* was similar to that in plantlets grown *in vitro* or was slightly increased, and a significantly increased uptake was found only in newly formed leaves (Deng and Donnelly, 1993). The above mentioned results suggest that photoinhibition might be the cause of the transient decrease in photosynthesis after transplantation.

#### *7.1.3.2. Humidity*

The retardation in development of cuticle, epicuticular waxes and functional stomatal apparatus during *in vitro* culture cause high stomatal and cuticular transpiration rates of leaves in plantlets when taken out of the culture vessels. In order to avoid this, slowly the plantlets should be transferred from high humidity to low humidity conditions. Micro shoots should be kept in the shade with plugs loosened and after a week or two, they should be transferred to pots containing sterile soil and sand mixture covered with polybags. Slowly stomatal and cuticular transpiration rates gradually decreases because stomatal regulation of water loss becomes more effective and cuticle and epicuticular waxes develop (Pospisilova *et al.*, 1999). Even if the water potential of the substrate is higher than the water potential of media with saccharides,the plantlets may quickly wilt as water loss of their leaves is not restricted.

In addition, water supply can be limiting because of poor conductivity of roots and root-stem connections (Fila *et al.*, 1998). Many plantlets die during this period. Short *et al.* (1987) reported that optimum growth and *in vitro* hardening of cultured cauliflower and chrysanthemum occurredwhen plantlets were cultured at 80% relative humidity. Leaves of chrysanthemum and sugar beet, which were initiated and developed at relative humidity below 100%, displayed increased epicuticular wax, improved stomatal functioning and reduced leaf dehydration (Ritchie *et al.*, 1991).

*7.1.3.3. Carbohydrate Concentration*

Plantlets are developed within the culture vessels under low level of light, on a medium containing ample sugar and nutrients to allow for heterotrophic growth and in an atmosphere with high level of humidity. Reports suggest that carbohydrate concentration influences the acclimatization process because plantlets switch from heterotrophic to autotrophic growth and any treatment before and after transfer increases the photosynthetic capacity of plants, which may improve plant establishment (Wainwright and Scrace, 1989). A sucrose concentration of 40 g/ladded prior to transferring water cress micro cutting to *in vivo* conditions was shown to increase the dryweight of established plantlets (Wainwright and Marsh, 1986). Bhatt and Dhar (2000) reported acclimatization of *Bauhinia vahlii* micro shoots. Rooted micro shoots, preconditioned in different sucrosesolution before transferring in the potting mixture, did not enhance percent survival but resulted in better quality of shoots. Shoot length, fresh, and dry weight of shoots were significantly better in 20 and 30 g/lsucrose as compared to control. Sucrose higher than 30 g/l reduced shoot growth of *ex vitro* established plants. Silvente and Trippi (1986) also reported that sucrose higher than 30 g/l was inhibitory to shoots of *Anagalis arvensis*.

### 7.1.4. Hardening in *In Vitro* Culture Condition

Deb and Imchen (2010) reported an alternative approach to acclimatize the micro-shoots of orchids by using an alternate substratum. Tissue cultured raised orchid seedlings are acclimatized and hardened *in vitro* by using 10% MS (Murashige and Skoog, 1962) basal medium with no carbon source or any plant growth regulators. In the culture vials, different types of matrix as an alternative substratum have been used for epiphytic and terrestrial orchids. A combination of charcoal pieces (5-7 mm in size), small brick chips and mosses at (1:1) ratio was found suitable for epiphytic orchids and (1:1) ratio of moss and decayed wood/

forest litter along with charcoal pieces, brick chips was preferred for terrestrial orchids. With the passage of time, the newly formed roots firmly attach to the charcoal chips. Similarly, Agnihotri *et al.* (2004) also reported 80% transplant success of *Carica papaya* plantlets when transferred from culture tubes along with soilrite plugs to the potted mix of garden soil and soil rite in the ratio of 1:1.

### 7.1.5. Antitranspirants and Growth Retardants in Humidity Control

Amaregouda *et al.* (1994) found that stomatal resistance was more in plants treated with 1500 ppm of Alar (B-9) and phenylmercuric acetate (PMA) (20 ppm), while Alachlor (20 ppm), Sunguard (0.02%), China clay (6% w/v) and silica powder (6% w/v)maintained moderate stomatal resistance compared to control. Testing of the film-forming antitranspirants like Aquawiltless, Clear spray, DC-200, Exhalt 4-10, Folicote, Protec, Vapor Gard and Wiltpruf forreduction of wilting on *Chrysanthemum morifolium* and *Dianthus caryophyllus* plantlets transferred *ex vitro*, showed that DC-200 had the greatest effect in reducing transpiration, but it had adverse effects onplant growth. All other anti-transpirants were ineffective in improving vigour of plants (Pospsilova *et al.*, 1999). Smith *et al.* (1990a&b) and Smith *et al.* (1991) reported that several growth retardants can be used in micro-propagation to reduce damage due to wilting without deleterious side effects. Use of paclobutrazol (0.5-4 mg/l) in the rooting medium is reported to result in reduced stomatal apertures, increased epicuticular wax, shortened stems and thickened roots, reduction in wilting after transfer to compost, andalso increased chlorophyll concentration per unit areaof leaf. Paclobutrazol was effective in inhibiting shoot growth and regulating various metabolic processeson apple trees. Steffens *et al.* (1985a&b) also reported that treatments with paclobutrazol resulted in a shift in the partitioning of assimilates from the leaves to the roots, increased carbohydrates in allparts of apple seedlings, chlorophyll, soluble proteinand mineral element concentration in leaf tissue, increased root respiration, reduced cell-wall polysaccharide and water loss, and accumulation of water stress-induced abscisic acid.

### 7.1.6. Abscisic Acid (ABA) and Ascorbic Acid in Acclimatization of Micro-shoots

Abscisic acid (ABA), a naturally occurring plant hormone critical for plant growth and development, plays an important role in plant water balance and in the adaptation of plants to stress environments including low temperature (Hetherington, 2001; Finkelstein and Gibson, 2002). It is transported via xylem to the shoot, where it regulates transpiration water loss and leaf growth (Hronkova *et al.*, 2003). Various stresses induce ABA synthesis and it is considered as a plant stress hormone (Tuteja, 2007). Adie *et al*. (2007) showed that ABA is an essential signal for plant resistance to pathogens affecting jasmonic acid (JA) biosynthesis and the activation of defences in *Arabidopsis*. Role of abscisic acid on tolerance to abiotic stresses has also been studied by Aguilar *et al.* (2000) in *Tagete serecta* in controlling leaf water loss, survival and growth of micro-shoots when transferred directly to the field. Hemavathi *et al.* (2010) found that ascorbic acid was expressed in transgenic potato by over-expressing L-gulonoc- lactone oxidase (GLOase) gene and accumulation of ascorbic acid showed tolerance to abiotic stresses.

### 7.1.7. Acclimatization for Biotic Stresses

Another major cause of high mortality of micro-shoots is their sudden exposure (particularly the root system) to microbial communities present in the soil as they do not possess sufficient resistance against the soil microflora. In the last few years, trials have been taken to expose the young *in vitro*-raised plantlets to useful microorganisms that promote growth and encourage mutual association. Endophyte refers to the fungi and bacteria which invade or live inside the tissues of plants without causing any diseases or injury to them. They also promote growth of the host plant and the formation of secondary metabolites related to plant defence (Hao *et al.*, 2010). Bio-hardening is an emerging dimension of micro-propagation techniques (Srivastava *et al.*, 2002).

*In vitro* co-culture of plant tissue explants with bacteria and vesicular arbuscularmycorrhiza induces developmental and metabolic changes in the derived plantlets which enhance their tolerance to abiotic and biotic stresses. The induced resistance response caused by the inoculants is referred to as 'biotization'. Hao *et al.* (2010) reported that treatment of suspension cells of *Ginkgo biloba* with fungal endophytes resulted in accumulation of flavonoids, increased ABA production and activation of phenylalanine ammonia- lyase. Bacteria associated with roots and rhizospheres of many plant species are also known as plant growth promoting rhizobacteria (Ramamoorthy *et al.*, 2001). Most of these bacteria belong to the genus *Bacillus* and *Pseudomonas* that proliferate around the tissues and also inside the plant tissues in a uniform environment free from fluctuations of temperature and moisture conditions. By using rhizosphere bacteria, successful biohardening of tissue cultured raised tea plants have been reported by Pandey *et al.* (2000). The bacterial isolates *Bacillus subtilis*, *Bacillus* sp. (associates of established tea rhizosphere), *Pseudomonas corrugata* 1 and *P.*

*corrugata* 2 (associates of young tea rhizosphere) were tested as microbial inoculants for hardening of tissue-cultured tea plants raised in the laboratory. The bacterial isolates colonized the soil rapidly and influenced the survival and growth of tea plants especially shoot length and leaf number in most cases. Trivedi and Pandey (2007) also reported inoculation of microshoots of *Picrorhiza kurrooa* by the three plant growth promoting rhizobacteria (*Bacillus megaterium*, *B. subtilis* and *Pseudomonas corrugata*) was effective in improving the survival of plants after transfer to soil. Maximum survival (94.5%) was observed in plants inoculated with *B. megaterium*. *Endophytic colonization* in rice (*Oryzasativa*L.) tissue cultured plants by using rhizobia has been studied by Senthilkumar *et al.* (2008). Calli treated with *Azorhizobium caulinodans* strains ORS571 and AA-SK-5 showed colonization in almost all parts of plant with significant increases in protein content, total nitrogen and nitrogenase activity. The regenerated plantlets showed a positive correlation in yield than uninoculated control. This also protects the juvenile axenic plants from infestation of the harmful saprophytes (Pandey *et al.*, 2000). Mycorrhization of tissue cultured plants provides advantage in terms of nutrient availability, soil pH, aeration and protection from water stress. Study on the micro-cloned plantlets of *Chlorophytum borivilianum* (Mathur *et al.*, 2008) showed more than 95% establishment in soil following treatment with various bioinoculants namely *Glomus aggregatum*, *Trichoderma harazianum* and *Piriformospora indica* whereas *Azospirullum* sp. (CIM-azo) and *Actinomycetes* sp. (CIM-actin) showed only up to 85% plantlet establishment. The un-rooted shoots also showed *in vivo* rooting (50%) when treated with mycorrhiza *Glomus aggregatum* (VAM) and *Trichoderma harazianum*.

## 7.2. METHOD OF PLANTLETS HARDENING

The transplanting process of *in vitro* grown plantlets can be a shock set out into *ex vitro* condition. These young plantlets can be made somewhat resistant to heat, cold temperatures, drying and whipping winds, certain types of insect injury and low soil moisture by a process termed hardening.The term hardening refers to any treatment that results in a firming or hardening of plant tissues. Such a treatment reduces the growth rate, thickens the cuticle and waxy layers. Such plants often have smaller and darker green leaves than non-hardened plants. Hardening results in an increased level of carbohydrates in the plant permitting a more rapid root development than occurs in non-hardened plants. There are different methods of hardening. Different methods and the different hardening media are being elaborated in this section.

### 7.2.1. Common Protocol for Hardening

The process of acclimatization of the *in vitro* grown plants to the normal environment in hardening micro-propagation has been extensively used for the rapid multiplication of many plants species. However, its wider use is restricted often by the high percentage of plant loss or damaged when transferred to *ex vitro* condition. The plants are subjected to specific culture regimes aimed at making them capable of surviving the uncontrolled and harsher *ex vitro* environment. The plants may be transferred to mist chamber or greenhouse and then to secondary hardening chambers for partial shade and further hardening to make it suitable for field plantation. The flowchart for hardening has been depicted in Fig. 7.1. Hardening of synthetic derived or tissue cultured derived plantlets is also being described stepwise in this section (Fig. 7.2).

After removing synthetic seed derived or*in vitro* derived plantlets from the containers, the gel like agar medium is gently washed from the roots

↓

Pre-hardened, synthetic seed derived or tissue cultured plantlets with well-developed roots becomes ready for planting into potting media in a greenhouse

↓

Pre-hardened plants should be quickly prepared and transferred for the greenhouse nursery under partial shade

↓

Plastic pots or polyethylene bags can be used as nursery containers. Poly-bags are preferred for their light weight

↓

Potting mixture with 2 parts growing media mixture, 1 part perlite, and 3 parts vermiculite sand is preferred for growing plantlets in greenhouse

↓

Slow-release fertilizers or liquid fertilizers are added to supply nutrient to plantlets. Multiple applications of liquid fertilizers are required

**Fig. 7.1.** Flow chart of hardening of regenerated plantlets

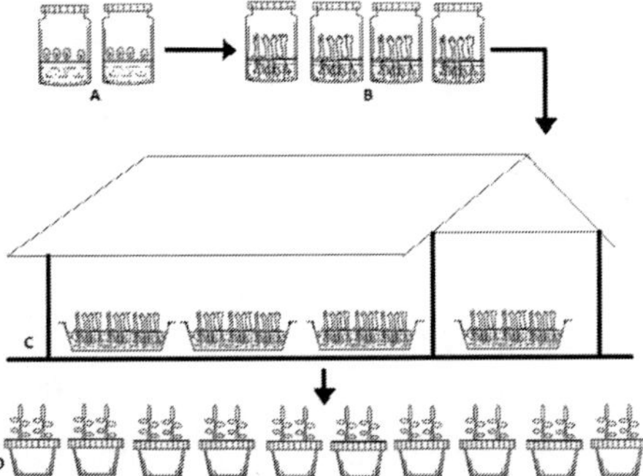


**Fig. 7.2.** Pictorial depiction of hardening of regenerated plantlets. **A)** Inoculation of synthetic seeds for germination; **B)** *In vitro* germinated synthetic seeds; **C)** *In vitro* grown plantlets are being hardened in net-house; **D)** *In vitro* hardened plantlets were taken under *in vivo* condition for final hardening.

#### *7.2.1.1. Dressing of well grown in vitro Plantlets*

Synthetic seeds derive plantlets or *in vitro* grown plantlets have synthetic medium adhering with roots and lower part of the plantlets. After removing plantlets from the containers, the gel like agar medium adhering with roots and lower part of the plantlets is to be gently washed from the roots.

#### *7.2.1.2. Potting Medium*

Potting medium is very important to initiate the growth of the plantlets under *in vivo* condition.Large number of potting media in different combinations and different proportion have been used for hardening of *in vitro* derived plantlets. Some of the potting media used for hardening have been briefly described in this section (Table 7.1).

**Table 7.1.** List of media used for hardening of tissue culture raised plantlets

| Media | Purpose | Remarks |
|---|---|---|
| Garden soil | Source of natural nutrients | To enhance water holding capacity |
| Sand | To enhance aeration | Need frequent watering and mineral nutrients |
| Peat | High water holding capacity and source of nitrogen | Partially decomposed water vegetation |
| Perlite | To enhance aeration | Grey/white material, contains no mineral nutrients |
| Vermiculite | Good source of magnesium and potassium | Magnesium, aluminium, iron-silicate (hydrated) |
| Soil rite mix | Good water and holding capacity enhance aeration | Mixture of peat, perlite and vermiculite |

**Garden soil:** The soil found in a typical yard will be about 90% mineral residue and only about 10% decayed organic matter. The term "garden soil" is often used on packages of premixed soils sold at home improvement stores, but it is not actually a separate type of soil. Garden soil is made up of decomposed matter, particularly the leaf litter. This layer is great for plants. Soil in a home garden has its own unique texture and combination of sand, silt, clay and various minerals. Packaged garden soils are mixed to incorporate a variety of soils and textures, and they are often mixed to target a particular type of garden or plant - you may find different mixes for flower gardens, vegetable gardens and even herb gardens. Garden soil mixture contains ingredients suitable for culturing tissue cultured derived plants in *in vivo*.

**Sand:** Sand has the largest particles and they are irregularly shaped. This is why sand feels course and also why it drains well. Sand does not compact easily.

Sand culture is a method of growing plants hydroponically, without the use of soil. It is a variation of gravel culture where the sand, which is used primarily to anchor plants in the grow bed or tray, is a lot finer than the gravel grow medium option. Sand culture is thought to be more efficient than traditional hydroponic methods because the sand largely decreases the risk of botanical ailments such as verticilium and fusarium. Because the plants receive fresh nutrient solutions after each watering cycle, they also tend to remain healthy, with no nutrient imbalances. According to botanists, sand also triggers a capillary action that encourages the nutrients to move laterally. This encourages an even distribution of nutrient solution across the roots. In addition, sand culture tends to be easier to maintain that other types of hydroponics. On the flip side, this soilless grown method sometimes requires additional agricultural pipes to channel away excess water, minerals, and nutrients.

**Peat:** It is a heterogeneous mixture of more or less decomposed plant (humus) material that has accumulated in a water-saturated environment and in the absence of oxygen. Its structure ranges from more or less decomposed plant remains to a fine amorphic, colloidal mass.It has high water holding capacity and source of nitrogen. Peat is sedentarily accumulated materials consisting of at least 30% (dry mass) of dead organic material.

**Perlite:** It is an amorphous volcanic glass that has relatively high water content, typically formed by the hydration of obsidian. It occurs naturally and has the unusual property of greatly expanding when heated sufficiently. Perlite can be used as a soil amendment or alone as a medium for hydroponics or for starting tissue cultured derived plants. When used as an amendment it has high permeability/low water retention and helps prevent soil compaction.

**Vermiculite:** It is a hydrous phyllosilicate mineral. It undergoes significant expansion when heated. It encourages seed germination and establishment of tissue cultured derived plants. It improves soil structure. This is because of its aeration and water-holding properties. It's a very light-weighted material, does not rot, improves soil structure through soil aeration and increases water and nutrient retention.

**Soilrite mix:** It is a mixture of 75% Irish peat moss and 25% horticultural grade expanded perlite having pH between 5.0 and 6.5. As soilrite mix contains peat moss with vermiculite, it increases water holding capacity and the proportion of peat is just to provide the plant with correct amount of air/ water ratio. Common features of the soilrite mix are:-

- It improves aeration and drainage
- It is free from disease and weed seeds/propagules
- As it is inorganic, it does not deteriorate
- pH essentially neutral (5.5 to 6.5)
- As it act as insulator, it reduces extreme soil temperature fluctuations

**Soil mixtures:** The soil should comprise of top soil/virgin soil + any of the following gravel, rice husks, sawdust, wood shavings, wood bark, or sand at a ratio of 6 parts to 1 part. Fertilizer application is not necessary at this stage.

**Soil sterilization:** It is important to sterilize soil to kill harmful organisms such as soilborne fungal diseases, nematodes and weeds. This can be done through either steam sterilization, soil solarization or chemical sterilization.

a) **Steam sterilization procedure:** This involves heating soil with a steam-air mixture. A simple steam sterilization kit is illustrated in figure 2. Two drums are required; the upright drum is filled with moist soil mixture. The horizontal one is half filled with water and then heated at 60-700 °C. The steam produced sterilizes the soil and then emerges through the top lid. After the steam emerges from the top of the drum, a thirty minutes (30) sterilization period is recommended. After which the heating can be stopped. Steam sterilization improves the soil physical structure, it is harmless to beneficial organisms in the soil, fast, easy, cheap, more effective than chemicals, and user friendly. The main disadvantage is the use of firewood which may be costly and environmentally unfriendly.

b) **Soil Solarization:** This involves using solar energy to heat the soil. The soil is collected and spread in a shallow 'pit' then covered with a clear polythene sheet in an open area to allow sunlight to penetrate. It takes 30-60 days depending on the weather.

c) **Chemical Sterilization:** This involves the use of chemicals. The commonly used chemicals are furadan (2 kg per tonne of moist soil); mocap 10G (3 kg per tonne) and Basamid. The soil is spread in a nursery-like bed, sprinkled with some water to moisten the soil. The chemical is mixed thoroughly with soil and the mixture covered with polythene. After one week the soil is turned over and covered again. After the second week, the soil is completely uncovered turned and left open for any leftover fumes to diffuse out. The soil is ready for use after three weeks.

### *7.2.1.3. Hardening Houses*

The plants are subjected to specific *ex vitro* condition aimed at making them capable of surviving the uncontrolled and harsher environment. The plants may be transferred to hardening houses, such as mist chamber or greenhouse and then to secondary hardening chambers for partial shade and further hardening to make it suitable for transplanting in the field conditions. Some of the hardening houses have been described in this section.

a) **Greenhouse**: A greenhouse (also called a glasshouse) is a structure with walls and roof made mainly of transparent material, such as glass, in which plants requiring regulated climatic conditions are grown. A more scientific definition is 'a covered structure that protects the plants from extensive external climatic conditions and diseases, creates optimal growth microenvironment, and offers a flexible solution for sustainable and efficient year-round cultivation (Redmond *et al.*, 2018)'. A modern greenhouse operates as a system.Therefore, it is also referred to as controlled environment agriculture, controlled environment plant production system, or phytomation system (Redmond *et al.*, 2018).

   Many commercial glass greenhouses or hothouses are hightech production facilities for vegetables or flowers. The glass greenhouses are filled with equipment including screening installations, heating, cooling, lighting, and may be controlled by a computer to optimize conditions for plant growth. Different techniques are then used to evaluate optimality-degrees and comfort ratio of greenhouse micro-climate (i.e., air temperature, relative humidity and vapor pressure deficit) in order to reduce production risk prior to cultivation of a specific crop (Shamshiri *et al.*, 2017).

b) **Net House:** Net house is basically naturally ventilatedclimate controlled. NetHouses arefor protecting plants from excess sunlight to stimulate the optimum plant growth.

c) **Shade House:** A shade house is a structure enclosed by agro nets or any other woven material to allow required sunlight, moisture and air to pass through the gaps. It creates an appropriate micro climate conducive to the plant growth. Shade nets are available in different shade percentage and range of colors like Green, Red, White and Black adjusting to different crops and a diversity of conditions,which helps in cultivation/ hardening of *in vitro* grown plants. It protects the plants against diseases, pests and natural weather disturbances.

d) **Polyhouse:** A polytunnel (also known as a polyhouse, hoop greenhouse or hoophouse, grown tunnel or high tunnel) is a tunnel typically made from steel and covered in polythene, usually semi-circular, square or elongated in shape. Polytunnels are mainly used in temperate regions in similar ways to glass greenhouses and row covers.

A hardening house should be located in a particular crop growing area. The hardening house should be easily accessible. Hardening house site should have sufficient water supply. The hardening house should be free from pests such as insects, fungi, nematodes, weevils. The hardening house should be fenced to protect seedlings from both wild and domestic damage.

### *7.2.1.4. Nursery Tools and Equipments*

Various tools are needed such as, potting bags, potting trays, polysheets, jembes, fork jembes, spades, rakes, trowels, buckets, wheel barrows, watering cans, hose pipes, and a store.

**Fig. 7.3.** Nursery tools and equipment. **A)** Potting bags; **B)** Potting trays; **C)** Polysheets; **D)** Spade; **E)** Fork jembes; **F)** rake; **G)** Trowel; **H)** Wheel barrow; **I)** Watering can.

### *7.2.1.4. Nursery Containers*

There are several types of nursery containers that can be used for hardening tissue culture derived plantlets, such as polyethylene plastic bags, clay pots, plastic pots, metallic pots, milk jugs, ice cream containers, bushel baskets, barrels, and planter boxes. Poly-bags are preferred for their light weight. These nursery containers are filled with potting mixture for transfer of tissue cultured derived plantlets for hardening.

Clay pots were the first types of containers used for nursery production. Plastic containers eventually replaced the clay pots and have become the industry standard as they are lightweight, inexpensive, and durable. Although plastic containers can be reused or recycled many times, they are not. With the move to increase sustainability in the Green Industries, many types of alternatives for petroleum-based plastic containers are being explored for use in container production. Alternative containers can generally be categorized into one of three groups: plantable, compostable, and recycled plastic/bio-based plastic containers, and are based on the degradation of the container after production and/ or landscape installation. An understanding of these characteristics is needed for best use and selection of alternative containers in production, retail, and landscape settings.

A challenge with plantable containers is that they are meant to degrade over time, which means that degradation is possible while plants are still in production or in a retail setting. Some plantable containers such as peat, manure, and some coir containers can have mold or algal growth that can be unappealing to consumers and may impact sales. Variable drying rates of substrates between container types should be taken into account in a retail setting to avoid over- and under- watering when container types are mixed in sale areas. Compostable containers can sometime look like plantable containers or even plastic containers, so it is important to understand what type of container you are working with. Plantable containers can be beneficial for landscapers due to time saving at installation from not removing containers and from saved cost of container disposal.

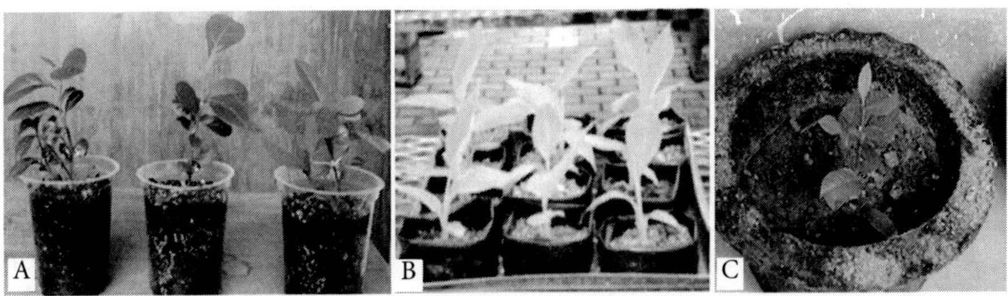


**Fig. 7.4.** Hardening containers. **A)** Plastic cups; **B)** Polyethylene bags; **C)** Earthen pot.

#### *7.2.1.5. Potting Mixture*

Type of potting mixture used during acclimatization is one of the important factors determining the survival percentage of the plants under *ex vitro* conditions. Potting mixture with two parts growing media mixture, 1 part Perlite, and 3 parts vermiculite sand is preferred for growing banana plantlets in greenhouse.

Parkhe *et al.* (2018) used different potting mixture such as garden soil + cocopeat (3:1), garden soil + farm yard manure (3:1), garden soil + vermicompost (3:1), garden soil + cocopeat + FYM+ vermicompost (2:1:1:1) and garden soil + sand + FYM + cocopeat (2:1:1:1). The *in vitro* rooted plantlets were hardened and acclimatized by using different treatments. Plantlets were transplanted from primary hardening after 45 days primary hardening gave maximum survival (100%) during transplanting in the field. These plants were hardened in polythene bags singly. The maximum survival during hardening (100%) was observed in shade net with maintained relative humidity and light intensity. Among these, the potting mixture containing garden soil and FYM (3:1) gave maximum height and survival of plantlets and shows outstanding performances in field condition.

#### *7.2.1.6. Microbial Inoculants for Hardening*

A successful attempt has been made to acclimatize the tissue cultured raised plants of *Tylophora indica* using various potting mixes of soil, vermicompost and biofertilizers (*Azotobacter* and *Pseudomonas*) in different combinations (Kaur *et al.*, 2011). A mixture of soil, vermicompost, *Azotobacter*, *Pseudomonas* in the ratio of 1, 1, 1 and 1, respectively showed the highest survival percentage upon transplantation of plants to the field conditions. *In vitro* raised plantlets of *Garciniaindica* showed 76% survival rate when hardened on cocopeat as a potting mixture (Chabukswar and Deodhar, 2005). Press mud cake mixed with soil was used as the optimal medium for producing sturdy plants during the secondary hardening process of banana plantlets (Vasane and Kothari, 2006). An efficient one-step hardening technique for tissue culture raised orchid seedlings was reported on chips of charcoal, bricks and decayed wood as an alternate substratum (Deb and Imchen, 2010). Vermicompost was shown to be the most suitable planting substrate for hardening which ensured high frequency survival (96%) of regenerated plants of *Tylophora indica* prior to their field transfer (Rani and Rana, 2010).

Four antagonistic bacterial isolates, *Bacillus subtilis*, *Bacillus* sp., *Pseudomonas corrugate* 1 and *P. corrugata* 2, isolated from the rhizosphere of tea plants growing in different geographical locations in India (Pandey *et al.*, 2000), were tested as microbial inoculants for hardening of tissue-cultured tea plants raised in the laboratory prior to the transfer to open land. Bacterial inoculations resulted in enhanced survival (up to 100, 96, and 88%), as against 50, 52, and 36% survival were observed in the corresponding control plants, in rainy, winter and summer seasons, respectively. Rhizoplane and rhizosphere soil analyses showed that the major biotic factor responsible for mortality following the transfer of tissue culture raised plants to soil was fungal attack (*Fusarium oxysporum*). Bacterial inoculations also resulted in plant growth promotion of tissue culture as well as seed raised plants of tea.

*7.2.1.7. Slow-Release Fertilizers or Liquid Fertilizers*

Slow-release fertilizers or liquid fertilizers are added to supply nutrient to plantlets. Multiple applications of liquid fertilizers are required.Slow-release fertilizers come in granules that look like beads and they are not water soluble the way liquid fertilizers are. They release gradual doses of nutrients into the soil, rather than flooding the soil with food all at once. Compared to liquid fertilizers, this makes them less likely to burn roots and foliage and gives them a longer duration. Depending on the formula, they can release nutrients for a few weeks or 8 to 9 months. Slow-release fertilizers are applied just once or twice a year, because they remain effective for longer than liquid fertilizers. Work can be done into the topsoil at planting or apply to the soil's surface later.

Liquid fertilizers act quickly to flush nutrients to plants. Initially, they can come as either a concentrate or a dry product, but both kinds are water soluble. Plant roots and leaves absorb them immediately when coming into contact with them, which makes them excellent if the plants are suffering from deficiencies or need a fast, heavy dose of nutrients. Liquid fertilizers and slow-release formulas require different methods of application in the garden or yard. When using liquid fertilizers, dilute them in water before applying them directly to the soil, close to the plant roots. To ensure, they maintain their effectiveness, reapply about 2 to 4 times a month.

Each type of fertilizer has unique benefits that can influence its decision on which works best for the garden. Concentrated liquid fertilizers are typically less expensive than dry, slow-release fertilizers. It can also be easier to apply, because the only need to go over the yard once, while slow-release fertilizers can require two or three passes over a yard to ensure full coverage, unless an efficient spreader is used. However, slow released fertilizers are often preferred for their long-term release of nutrients, meaning there is no need to reapply them as often as liquid fertilizers is done.

### *7.2.1.8. Transfer to Main Field*

Young, pampered *in vitro* grown plantlets that were grown either indoors or in a greenhouse /net house/ under shade net will need a period to adjust and acclimate to outdoor conditions, prior to planting in the garden. This transition period is called 'hardening off'. Hardening off gradually exposes the tender plants to the wind, the sun, and rain and toughens them up by thickening the cuticle on the leaves so that the leaves lose less water when exposed to the elements. This helps prevent transplant shock; seedlings that languish, become stunted or die from sudden changes in temperature. The length of time a seedling requires to harden off depends on the type of plants and the temperature and temperature fluctuations.

Nursery containers into which the plantlets will be transplanted are prepared. Nursery containers containing sterile soil mixture are used for transplanting of tissue cultured derived plantlets. Rooted plantlets are removed from the growth chamber. Nursery containers are filled with sterile soil.

i. Plantlets are carefully removed from jars or culture tubes.
ii. The medium adhering to the base of the plantlets is washed off gently without any injuring.
iii. The plantlets are taken to the hardening nursery and shed as soon as possible preferably on the same day.
iv. They are transplanted into seedling trays or tunnels in rows and watered twice or once a day depending on weather conditions.
v. One month later, they are transferred into 6 × 9 polythene bags containing sterilized soil and manure.
vi. Watering continues as necessary.

The seedlings are monitored regularly for somaclonal variants. Any plants showing unusual growth habits are removed and kept aside for further observation. Some varieties may show some purplish coloration on the leaves which is normal in the early stages of growth. The seedlings are ready for field transplanting after two months when they are about one foot tall and have at least five leaves.

**Transfer and Establishment of Tissue Cultured derived Banana Plants:** Banana plants are allowed to acclimatize in greenhouse for two months with plant height up to 20 cm, before they get ready to transplant in the field. Young plants typically produce a new leaf approximately every five days during their early development. Gestation and stabilization periods of seedlings depend on nutrient status of the soil and type of cultivar and it ranges from 9 to10 month after transplanting.

## 7.2.2. Nursery Hygiene and Organization

i. The nursery should be kept clean and free from weeds.

ii. Pests and diseases should be controlled immediately after they are spotted.

iii. Seedlings should be arranged in neat rows with clear paths for walking and transportation.

iv. Varieties should be separated and labeled accordingly, showing information such as date of transplanting and number of seedlings for ease of marketing

## References


Adie BAT, Perez-Perez J, Perez-Perez MM Godoy M, Sánchez-Serrano JJ, Schmelz EA, Solano R. 2007. ABA is an essential signal for plant resistance to pathogens affecting JA biosynthesis and the activation of defenses in *Arabidopsis*. Plant Cell. 19: 1665-1681

Agnihotri S, Singh SK, Jain M, Sharma M, Sharma AK, Chaturvedi HC. 2004. *In vitro* cloning offe male and male Carica papaya through tips of shoots and inflorescences. Indian J Biotechnol. 3: 235-240.

Aguilar ML, Espadas FL, Coello J, Maust BE, Trejo C, Robert ML, Santamaría JM. 2000. The role of abscisic acid in controlling leaf water loss, survival and growth of micropropagatedTageteserecta plant when transferred directly to the field. J Exp Bot. 51: 1861-1866.

Amaregouda A, Chetti MB, Salimath PM,Kulkarni SS. 1994. Effect of antitranspirants on stomatal resistance and frequency, relative water content and pod yield in summer groundnut (*Arachishypogaea* L.). Ann Plant Physiol. 8: 18-23.

Baroja Fernandez ME.1993.RelacionesHidricas y ActividadFotosintetica de Plantas de PatataCultivadas in Vitro. [Water Relations and Photosynthetic Activity of Potato Plantlets Cultivated in Vitro.], Thesis, University of Navarra, Pamplona.

Baroja ME, Aguirreolea J, Sánchez-Díaz M. 1995.$CO_2$ exchange of in vitro and acclimatizated potatoplantlets. In: Carre F, Chagvardieff P (ed): Ecophysiology and Photosynthetic In Vitro Cultures. CEA, Centre d'Études de Cadarache, Saint-Paul-lez-Durance. pp. 187-188.

Bhatt ID, Dhar U. 2000. Combined effect of cytokinins on multiple shoot production from cotyledonary node explants of Bauhinia vahlii. Plant Cell Tissue Org Cult. 62: 79-83.

Brainerd KE, Fuchigami LH, Kwiatkowski S, Clark CS. 1981. Leaf anatomy and water stress of aseptically cultured "pixy" plum grown under different environments. HortScience.16: 173-175.

Chabukswar MM, Deodhar A Manjushri.2005. Rooting and hardening of *in vitro* plantlets of *Garciniaindica* Chois. Indian J Biotechnol.4: 409-413.

Dami I, Hughes H. 1995.Leaf anatomy and water loss of in vitro PEG-treated "Valiant" grape. Plant Cell Tissue Organ Cult.42: 179-184.

Deb CR, Imchen T. 2010. An efficient *in vitro* hardening of tissue culture raised plants. Biotechnol. 9: 79-83.

Deng R, Donnelly J. 1993. *In vitro* hardening of red raspberry through $CO_2$ enrichment and relative humidity reduction on sugar-free medium. Can J Plant Sci. 73: 1105-1113.

Díaz-Pérez JC, Sutter EG, Shackel KA. 1995. Acclimatization and subsequent gas exchange, water relations, survival and growth of microcultured apple plantlets after transplanting them in soil. Physiol Plant. 95: 225-232.

Diettrich B, Mertinat H, Luckner M. 1992. Reduction of water loss during ex vitro acclimatization of micropropagated *Digitalis lanata* clone plants. Biochem Physiol Pflanz. 188: 23-31.

Donnelly DJ, Vidaver WE. 1984. Leaf anatomy of red raspberry transferred from culture to soil. J AmerSocHort Sci. 109: 172-176.

Drew AP, Kavanagh KL, Maynard CA. 1992.Acclimatizing micropropagated black cherry by comparison with half-sib seedlings. Physiol Plant. 86: 459-464.

Fila G, Ghashghaie J, Hoarau J, Cornic G. 1998. Photosynthesis, leaf conductance and water relations of in vitro cultured grapevine rootstock in relation to acclimatisation. Physiol Plant. 102: 411-418.

Finkelstein RR, Gibson SI. 2002. ABA and sugar interactions regulating development: cross-talk or voices in a crowd? CurrOpin Plant Biol. 5: 26-32.

Gilly C, Rohr R, Chamel A. 1997.Ultrastructure and radiolabelling of leaf cuticles from ivy (Hedera helix L.) plants in vitro and during ex vitro acclimatization. Ann Bot. 80: 139-145.

Grout BWW, Aston MJ. 1977. Transplanting of cauliflower plants regenerated from meristem culture. I. Water loss and water transfer related to changes in leaf wax and to xylem regeneration. Hort Res. 17: 1-7.

Grout BWW, Aston MJ. 1978. Modified leaf anatomy of cauliflower plantlets regenerated from meristem culture. Ann Bot. 42: 993-995.

Hao G, Du X, Zhao F, Ji H. 2010. Fungal endophytes-induced abscisic acid is required for flavonoid accumulation in suspension cells of Ginkgo biloba. Biotechnol Lett. 32: 305-314.

Hazarika BN. 2003. Acclimatization of tissue-cultured plants. Curr Sci. 85: 1704-1712.

Hemavathi A, Upadhyaya CP, Akula N, Young KE, Chun SC, Kim DH, Park SW. 2010. Enhanced ascorbic acid accumulation in transgenic potato confers tolerance to various abiotic stresses. BiotechnolLett. 32: 321-330.

Hetherington AM. 2001. Guard cell signalling. Cell. 107: 711-714.

Hronkova M, Zahradnickova H, Simkova M, Šimek P, Heydová A. 2003. The role of abscisic acid in acclimation of plants cultivated in vitro to *ex vitro* conditions. Biol Plant 46: 535-541.

Johansson M, Kronestedt-Robards EC, Robards AW. 1992. Rose leaf structure in relation to different stages of micropropagation. Protoplasma.166:165-176.

Kadlecek P. 1997. [Effect of pretreatment by irradiance and exogenous saccharose under in vitro conditions on photosynthesis and growth of tobacco (*Nicotianatabacum* L.) plants during acclimatization after transfer to soil.], Diploma Thesis, Charles University, Department of Plant Physiology, Praha.

Kaur H, Anand M, Goyal D. 2011. Optimization of potting mixture for hardening of in vitro raised plants of Tylophoraindica to ensure high survival percentage. Int J Med Arom Plants. 1(2): 83-88.

Lavanya M, Venkateshwarlu B, Devi BP. 2009. ***Acclimatization of Neem Microshoots*** adaptable to semi-sterile conditions.Indian J Biotechnol. 8: 218-222.

Lee N, Wetzstein HY, Sommer HE. 1985. Effects of quantum flux density on photosynthesis and chloroplast ultrastructure in tissue-cultured plantlets and seedlings of Liquidambar styraciflua L. towards improved acclimatization and field survival. Plant Physiol. 78: 637-641.

Lee N, Wetzstein HY, Sommer HE. 1988. Quantum flux density effects on the anatomy and surface morphology of in vitro- and in vivo-developed sweetgum leaves. J AmerSocHort Sci. 113: 167-171.

Marin JA, Gella R, Herrero M. 1988. Stomatal structure and functioning as a response to environmental changes in acclimatized micropropagated *Prunuscerasus* L. Ann Bot. 62: 663-670.

Mathur A, Mathur AK, Verma P, Yadav S, Gupta ML, Darokar MP. 2008. Biological hardening and genetic fidelity testing of micro-cloned progeny of Chlorophytum borivilianum. Afr J Biotechnol. 7: 1046-1053.

Murashige T, Skoog F. 1962. A revised media for rapid growthan d bioassays with tobacco tissue cultures. Physiol Plant. 15: 473-497.

Noé N, Bonini L. 1996.Leaf anatomy of highbush blueberry grown in vitro and during acclimatization to ex vitro conditions.Biol Plant. 38: 19-25.

Pandey A, Palni LMS, Bag N. 2000. Biological hardening of tissue culture raised tea plants through rhizosphere bacteria. Biotechnol Lett. 22: 1087-1091.

Parkhe S, Dahale M, Ningot E, Gawande S. 2018. Study of Different Potting Mixture on Hardening of Banana Tissue Culture Plantlets Its Field Performance. Int J Curr Microbiol App Sci. 6: 1941-1947.

Pospíšilová J, Synková H, Haisel D, Catský J, Wilhelmová N, Šrámek F. 1999.Effect of elevated $CO_2$ concentration on acclimation of tobacco plantlets to ex vitro conditions. J Exp Bot. 50: 119-126.

Pospisilova J, Ticha I, Kadlecek P, Haisel D, Plzáková S. 1999. Acclimatization of micropropagated plants to ex vitro conditions. Biol Plant. 42: 481-497.

Pospíšilová J, Wilhelmová N, Synková H, Catsky J, Krebs D, Tichá I, Hanácková B, Snopek J. 1998. Acclimation of tobacco plantlets to ex vitro conditions as affected by application of abscisic acid. J Exp Bot. 49: 863-869.

Preece JE, Sutter EG. 1991. Acclimatization of micropropagated plants to the greenhouse and field. In: Debergh PC, Zimmerman RH (eds.) Micropropagation. Technology and application.Kluwer Academic Publishers, Dordrecht. pp. 71-93.

Ramamoorthy V, Vishwanathan R, Raguchander T, Prakasam V, SamiyappanR. 2001. Induction of systematic resistance by plant growth promotingrhizobacteria in crop plants against pests and diseases. Crop Prot. 20: 1-11.

Rani S, Rana JS. 2010. *In vitro* Propagationof *Tylophoraindica*-Influence of ExplantingSeason, Growth Regulator Synergy, Culture Passage and Planting Substrate. J American Sci.6(12): 385-392.

Redmond S, Fatemeh K, Ting CK, Kelly T, Ibrahim H, Cornelia W, Desa A; Zahra S. 2018. Advances in greenhouse automation and controlled environment agriculture: A transition to plant factories and urban agricultur. Interl J AgriculBiol Engineering. 11. doi:10.25165/j.ijabe.20181101.3210 – via Research Gate.

Ritchie GA, Short KC, Davey MR. 1991.*In vitro*acclimatizationof chrysanthemum and sugar beet by treatment with paclobutrazol and exposure to reduced humidity. J Exp Bot. 42: 1557-1563.

Rival A, Beulé T, Lavergne D, Nato A, Havaux M, Puard M. 1997.Development of photosynthetic characteristics in oil palm during in vitro micropropagation. J Plant Physiol. 150: 520-527.

Sallanon H, Berger M, Genoud C, Coudret A. 1998 Water stress and photoinhibition in acclimatization of Rosa hybrida plantlets. In Vitro Cell Dev Biol. 34: 169-172.

Senthilkumar M, Madhaiyan M, Sundaram SP et al (2008) Induction of endophytic colonization in rice (*Oryzasativa* L.) tissue culture plants by *Azorhizobiumcaulinodans*. Biotechnol Lett 30:1477–1487

Shamshiri J. 2017. Dynamic Assessment of Air Temperature for Tomato (Lycopersiconesculentum Mill) Cultivation in a Naturally Ventilated Net-Screen Greenhouse under Tropical Lowlands Climate. J AgrilSci Technol. 19(1): 59-72.

Short KC, Warburton J, Roberts AV (1987) In vitro hardening of cultured cauliflower and chrysanthemum to humidity. ActaHortic 212:329–340

Short KC, Wardle K, Grout BW, Simpkins I. 1984. In vitro physiology and acclimatization of asseptically cultured plantlets. In: Novák FJ, Havel L, Doležel J (ed): Plant Tissue and Cell Culture Application to Crop Improvement. Inst Exp Bot CzechoslovAcadSci, Prague. pp. 475-486.

Silvente S, Trippi VS (1986) Sucrose modulated morphogenesis in *Anagalisarvensis* L. Plant Cell Physiol 27:249–354

Singsit C, Veilleux RE. 1991. Chloroplast density in guard cells of leaves of anther-derived potato plants grown in vitro and in vivo. HortScience.26: 592-594.

Smith EF, Roberts AV, Mottley J (1990a) The preparation in vitro of chrysanthemum for transplantation to soil. 2. Improved resistance to desiccation conferred by paclobutrazol. Plant Cell Tissue Org Cult 21:133–140

Smith EF, Roberts AV, Mottley J (1990b) The preparation in vitro of chrysanthemum for transplantation to soil 3. Improved resistance to desiccation conferred by reduced humidity. Plant Cell Tissue Org Cult 21:141–145

Smith EF, Roberts AV, Mottley J et al (1991) The preparation in vitro of chrysanthemum for transplantation to soil. 4. The effect of eleven growth retardants on wilting. Plant Cell Tissue Org Cult 27:309–313

Srivastava PS, Bharti N, Pandey D et al (2002) Role of mycorrhiza in in vitro micropropagation of plants. In: Mukerji KG, Manoharachari C, Chamola BP (eds) Techniques in mycorrhizal studies. Kluwer Academic Publishers, The Netherlands, pp 443–460

Steffens GL, Wang SY, Faust M et al (1985a) Controlling plant growth via the gibberellin biosynthesis system I growth parameter alterations in apple seedlings. Physiol Plant 63:163–168

Steffens L, Wang SY, Faust M et al (1985b) Growth carbohydrate and mineral element status of shoot and spur leaves and fruit of 'Spartan' apple trees treated with paclobutrazol. J Am SocHorticSci 110:850–855

Sutter E. 1988.Stomatal and cuticular water loss from apple, cherry, and sweetgum plants after removal from in vitro culture. J AmerSocHort Sci. 113: 234-238.

Synková H. 1997. Sucrose affects the photosynthetic apparatus and the acclimation of transgenic tobacco to ex vitro culture. Photosynthetica.33: 403-412.

Tichá I, Kutík J. 1992.Leaf morphology, anatomy and ultrastructure in transgenic tobacco (*Nicotianatabacum* L. cv. Samsun) plantlets. Biol Plant. 34: 567-568.

Ticha I, Radochova B, Kadlecek P. 1999. Stomatal morphology during acclimatization of tobacco plantlets to ex vitro conditions. Biol Plant. 42: 469-474.

Trillas MI, Serret MD, Jorba J, Araus JL. 1995. Leaf chlorophyll fluorescence changes during acclimatization of the rootstock GF677 (peach × almond) and propagation of Gardenia jasminoides E. In: Carre F, Chagvardieff P (ed): Ecophysiology and Photosynthetic In Vitro Cultures. CEA, Centre d'Études de Cadarache, Saint-Paul-lez-Durance. pp. 161-168.

Trivedi P, Pandey A. 2007. Biological hardening of micropropagated Picrorhizakurrooa Royel ex Benth., an endangered species of medical importance. World J MicrobiolBiotechnol. 23: 877-878

Tuteja N. 2007. Abscisic acid and abiotic stress signaling.Plant Signal Behav. 2: 135-138.

Van Huylenbroeck JM, Debergh PC. 1996. Impact of sugar concentration in vitro on photosynthesis and carbon metabolism during ex vitro acclimatization of Spathiphyllum plantlets. Physiol Plant. 96: 298-304.

Van Huylenbroeck JM, Huygens H, Debergh PC. 1995. Photoinhibition during acclimatization of micropropagatedSpathiphyllum "Petite" plantlets. *In Vitro* Cell Dev Biol. 31: 160-164.

Van Huylenbroeck JM, Piqueras A, Debergh PC.1998 Photosynthesis and carbon metabolism in leaves formed prior and during ex vitro acclimatization of micropropagated plants. Plant Sci. 134: 21-30.

Van Huylenbroeck JM. 1994. Influence of light stress during the acclimation of in vitro plantlets. In: Struik PC, Vredenberg WJ, Renkema JA, Parlevliet JE (eds.) Plant Production on the Threshold of a New Century. Kluwer Academic Publishers, Dordrecht, NL. pp. 451-453.

Vasane RS, Kothari RM. 2006.Optimizationofhardening process of banana plantlets*(Musa paradisiaca*L. var grand nain). IndianJ Biotechnol.5: 394-399.

Wainwright H, Marsh J. 1986.The micropropagation ofwatercress (*Rorippa nasturtium-*aquaticum L.). J HorticSci. 61:251-256.

Wainwright H, Scrace J. 1989. Influence of *in vitro*preconditionin with carbohydrates during the rooting of microcuttings on *in vivo* establishment. SciHortic. 8: 261-266.

Waldenmaier S, Schmidt G. 1990.HistologischeUnterschiedezwischen in-vitro- und ex-vitro-Blätternbei der Abhärtung von Rhododendron.Gartenbauwissenschaft.55: 49-54.

Wetzstein HY, Sommer HE. 1982. Leaf anatomy of tissue-cultured Liquidambar styraciflua (*Hamamelidaceae*) during acclimatization. Amer J Bot. 69: 1579-1586.

Wetzstein HY, Sommer HE. 1983. Scanning electron microscopy of in vitro-cultured Liquidambar styraciflua plantlets during acclimatization. J AmerSocHort Sci. 108: 475-480.

Zacchini M, Morini S. 1998.Stomatal functioning in relation to leaf age in in-vitro-grown plum shoots. Plant Cell Rep. 18: 292-296.

# 8

# Achievements

Many researchers successfully prepared synthetic seeds of crop plants. The major success of this technology is being elaborated in this chapter.

## 8.1. FIELD CROPS

Rice is the world's most important food crop and a primary food source for more than one third of the world's population. This crop has received considerable attention in biotechnology research progammes. Research in artificial seeds in rice is not scanty and this technology through somatic embryogenesis would offer a great scope for large-scale propagation of superior elite hybrids (Brar and Khush, 1994). However, success on artificial seeds research in field crops, especially cereals is still in infancy. However, many workers successfully encapsulated somatic and androgenic embryos of cereal crops (Datta and Potrykus, 1989; Suprasanna *et al.,* 1996; Roy and Mandal, 2004 & 2008). Brar *et al.* (1994) emphasized the need for research on artificial seeds in rice through embryogenesis and outlined its impact on mass propagation of true-breeding hybrids. Somatic embryogenesis can also be used in the regeneration of genetically transformed plants (Vicient and Martinez, 1998). Subsequently those somatic embryos could be economically and successfully propagated though synthetic seed technology.

Somatic embryos of rice have been encapsulated to produce synthetic seeds (Suprasanna *et al.,* 1996). The encapsulated embryos showed better conversion into plantlets than non-encapsulated embryos. Xing *et al.* (1995) prepared artificial seed from hybrid embryos of *japonica* and *javanica* rice. The germination rate of artificial seeds in sterile conditions was 15-60% on vermiculite and agar media. The induction of pollen embryogenesis genetically differs from zygotic embryogenesis, and the androgenic embryos, may be used for synthetic seed production. Roy and Mandal (2004, and 2008) prepared synthetic seeds from androgenic embryos, embryo-like structures and microtillers of *indica* rice var. IR 72 (Fig. 5.10). The results indicated higher germination in the beaded embryos than the unbeaded embryos. The reduced rate of germination of artificial seeds may be attributed to the damage incurred while separating the embryos from clusters and/or owing to adverse effects of chemicals used for encapsulation. Jaseela *et al.* (2009) successfully encapsulated the somatic embryos of *Oryza nivara*, wild species of rice. Datta and Potrykus (1989) encapsulated microspore-derived embryos of barley and germination response was found to be high (80%) and seedlings were more vigourous than that of non-encapsulated embryos. However, further research is needed to optimize protocols for production of androgenic viable synthetic seeds of cereal crops.

## 8.2. VEGETABLE CROPS

To scale up the production system and encapsulation technologies production of encapsulated units were developed as a technology for transport using somatic embryos of crop plants. Development of insect resistant crop plant by genetic engineering can be an alternative measure to combat the problem. But in this case maintenance of resultant genotype (for example, hybrid, meiotically unstable genotypes and genetically engineered genotypes) may be a great problem or inhibitor. Most of the vegetable crops are highly cross pollinated, thus a variety of vegetable consists heterogeneous population. Artificial seed in vegetable can offer a new avenue for supporting the programme of genetic engineering providing an exciting asexual propagation bridge for readily multiplication of transformed plants. Encapsulated carrot artificial seeds showed 97% germination (Sakamato *et al.*, 1992). Onishi *et al.* (1994) also encapsulated carrot somatic embryos showed 52% conversion after sowing on a humid soil in green house. Takahata *et al.* (1992) produced artificial seeds of Chinese cabbage, broccoli and rape following desiccation of embryos. Huda and Bari (2007) prepared synthetic seeds of eggplant which could be stored at 4 °C temperature for 45 days.

Celery is an important vegetable of temperate zone. Many authors (Kim and Janick, 1990; Hayashi, 1993; Onishi *et al.*, 1994; Suchara *et al.*, 1995) encapsulated somatic embryos of celery to produce synthetic seeds. Hayashi (1993) produced dehydrated high quality somatic embryos of $F_1$ celery (*Apium graveolens*), which were coated with an alginate gel and artificial endosperm, using encapsulating machines. Kim and Janick (1990) also prepared synthetic seeds of celery by desiccation. Hydrated synthetic seeds were prepared from *in vitro* produced embryoids of $F_1$ hybrids of celery with alginic acid by Sakamoto *et al.* (1991). They found that increasing in the thickness of the coating delayed development of the synthetic seed.

A high number of micro-shoots of *Brassica oleracea* var. *botrytis* (cauliflower) were obtained by (Siong *et al.*, 2012) when hypocotyl explants from 2-week-old aseptic seedlings were cultured on MS medium supplemented with 0.1 mg/L NAA and 5 mg/L BAP. Artificial or synthetic seeds were formed when the micro shoots were encapsulated in 4% (w/v) sodium alginate with 100 mM $CaCl_2$ as complexing solution. The artificial seeds took 12 days and 14 days to germinate on MS basal medium. The germination percentage of artificial seeds was enhanced by the inclusion of 0.3 mg/L NAA and 3.0 mg/L BAP in the encapsulation matrix after 7 and 30 days of pregermination storage. The time taken for germination was also faster when MS fortified with 0.3 mg/L NAA and 3.0mg/L BAP were used. Isolated micro shoots encapsulated in MS supplemented with 0.3 mg/L NAA and 3.0mg/L BAP gave high germination percentages after 7 and 30 days of pre-germination storage period, respectively. Complete plant regeneration was achieved from the germination of these artificial seeds derived from the micro shoots after 3-4 weeks in culture.

*Ipomea aquatica* nodes with axillary buds were encapsulated with sodium alginate (Tang *et al.*, 1994a). The percentage germination of the artificial seeds was 42, 26 and 18% after low temperature storage (4 °C) for 30, 60, 90 days, respectively. Tang *et al.* (1994b) prepared artificial seeds with 4% sodium alginate. They found that germination of synthetic seeds of sweet potato increased with increasing length of axillary bud. Germination of artificial seeds and number of seedlings produced were higher on MS agar medium than on humus. Seed derived bipolar somatic embryos of *Solanum melongena* cv. Giulietta were encapsulated in sodium alginate (Mariani, 1992). Gibberellic acid ($GA_3$) was added to the sodium alginate. Plantlet germination rate was 67% when artificial seeds were obtained using Na-alginate encapsulation without $GA_3$. Sucrose and $GA_3$ were found to have negative effects on synthetic seed germination.

## 8.3. FRUIT CROPS

Most of the fruit trees are perennial and have to pass through a long juvenile period for fruit initiation. Micropropagation via encapsulation of micropropagating materials will be beneficial to reduce the juvenile period as well as the large-scale production of true-to-type planting materials. Ara *et al.* (1999) encapsulated individual embryos originating from nuclear explants of *Mangifera indica* L. cv. Amarapalli in 2% algenic acid-sodium salt gel. They observed that the percentage of germination of encapsulated somatic embryos was higher than that of naked somatic embryos of the same size on the same medium. Well-developed plantlets regenerated from encapsulated somatic embryos have been successfully established in soil. Transfer of plants, obtained from somatic embryos of some cultivars of mango, to soil has also been reported by Litz *et al.* (1993).

Somatic embryos of *Citrus reticulata* encapsulated with an artificial endosperm containing $GA_3$; frequency of conversion of somatic embryos on plant growth medium was 50% (Antonietta *et al.*, 1998). Encapsulation with $GA_3$ was also found to be useful for storage of somatic embryos at 4 °C for 1 month. Nieves *et al.* (1998) also successfully encapsulated zygotic embryos of *Citrus reticulata.* Madhav *et al.* (2002) developed protocol for nuclear embryogenesis (technique for producing virus-free plantlets) and artificial seed preparation in *C. reticulata* cv. Kharimandarin. Upon subculture, the artificial seeds differentiated into whole plants *in vitro*. Apple rootstock M.26 had been successfully propagated through encapsulated shoot tips (Sicurani *et al.*, 2001; Brischia *et al.*, 2002). The results demonstrated that synthetic seeds of M.26 apple root stock can be produced through organogenesis followed by root induction and encapsulation of differentiating propagule.

Encapsulated buds excised from *in vitro* proliferated shoots of Hayward Kiwi fruits were used for non-embryogenic synthetic seed production (Andriani *et al.*, 2000). Conversion of those synthetic seeds into plantlets was 57.5%. Non-embryogenic synthetic seeds were developed by encapsulating microcuttings derived from *in vitro* proliferated shoots of kiwi fruit cv. Top star (Alvarez *et al.*, 2002). Whole plantlets conversion of synthetic seeds was found in between 45 and 85% based on the substrate on which the synthetic seeds were sown. Somatic embryos from shoot tip culture of Zaghloul Date Palm were successfully encased in sodium alginate capsules by Bekheet *et al.* (2002).

Priya *et al.* (2003) established a protocol to obtain synthetic seed production in banana cv. Poovan. Synthetic seeds of *Carica papaya* hybrid Yuan You 1 showed 80% germination in hormone free medium *in vitro* and 32% in sterilized soil (Ye *et al.*, 1993). The survival rate of seedling reached to 72% after which they were transplanted into an experimental field. Plants derived from artificial seeds grew normally to flowering and set fruit. They observed no variation in either morphological characters or isozymatic patterns of peroxidase, esterase and amylase.

## 8.4. ORNAMENTAL CROPS

Ornamental plants are particularly suitable for propagation through artificial seed because of the high turnover in the market and difficulties faced in producing true seeds. Ruffoni (1998) elaborated the importance and procedure of artificial seeds production of ornamental plants from suspension culture. Ruffoni *et al.* (1994) investigated on the potentiality of artificial seed production of Lizianthus (*Eustoma grandiflorum*) and *Genista monosperma* to improve the germination and establishment of the embryos. They found that the addition of Zeatin (0.5 mg/L) to the alginate encapsulation coating improved the shoot production and sucrose (40 g/L) improved the percentage of root emergence of *Lizianthus*.

### 8.4.1. Orchids

Artificial seed technology has been well explored in orchid research. Artificial seeds were formed by the encapsulation of *Cymbidium longifolium* protocorms (Singh, 1988). *Cymbidium* protocorm- like body (PLB) Artificial seed embedded in a fungicide, and cocooned in chitosan resulted in a 35% germination rate when sown directly on non-sterilized substrate (Nhut *et al*., 2005). Protocorms of 3-4 mm in size were shown to be suitable for optimal conversion frequency of encapsulated PLBs of *Cymbidium giganteum*, smaller PLBs not able to withstand encapsulation or requiring a long time to emerge out of the capsule (Corrie and Tandon, 1993). High germination frequencies of *C. giganteum* encapsulated PLBs (Saiprasad and Polisetty, 2003) or *C. longifolium* protocorms (Chetia *et al*., 1998) occurred when they were stored

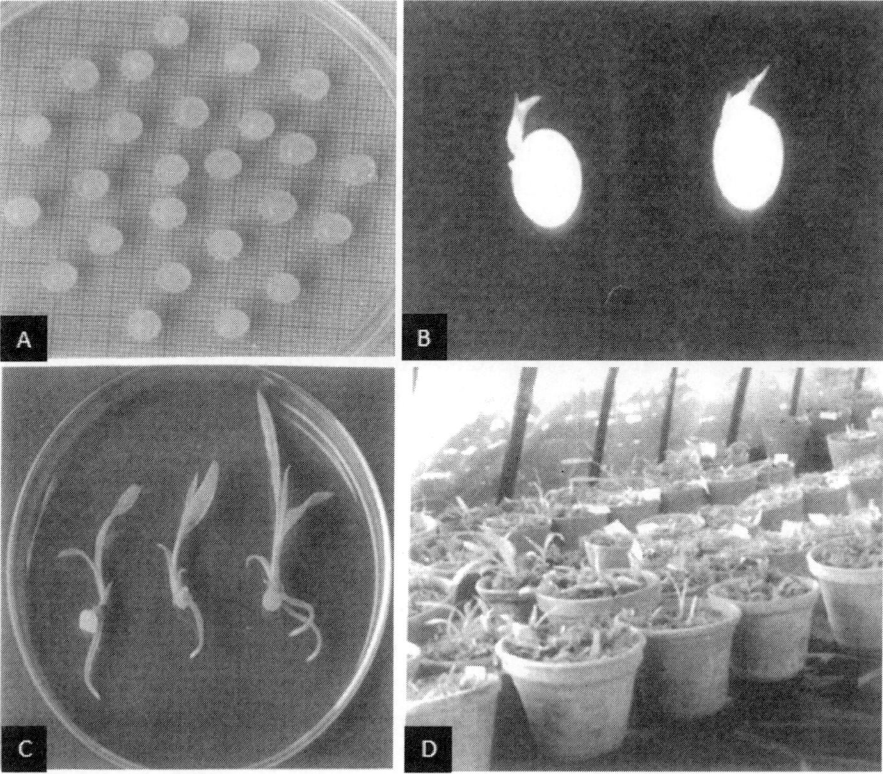


**Fig. 8.1.** Pictorial depiction of synthetic seed production and plantlet regeneration in *Orchid* sp. **A)** Encapsulated protochorm in sodium alginate beads; **B)** Germinating synthetic seeds *in vitro*; **C)** Seedling developed from synthetic seeds with roots; **D)** Plantlets under glasshouse condition for hardening.

at 4 °C. In the former study, 3% sodium alginate upon complexion with 75 mM calcium chloride for 20-30 min was the optimum combination for proper hardening of beads viz., *Dendrobium*, *Oncidium* and *Cattleya* orchids (Saiprasad and Polisetty, 2003).

Datta *et al.* (1999) produced artificial seeds through encapsulation of protocorm-like bodies (PLBs) of *Geodorum densiflorum* (Lam) Sehltr, an endangered orchid from the Terai Hills, North-eastern Himalaya. Many other workers also successfully prepared synthetic seeds of orchids using PLBs (Bhattacharjee *et al.*, 1998; Sharma *et al.*, 1992; Malemnganba *et al.*, 1996). Feshly encapsulated protocorms of three orchid species- *Arides odoratum*, *P. takervillae* and *Rhyncostylis retensa* showed more than 90% regeneration (Fig. 8.1). Synthetic seed were produced from protocorm-like bodies (PLBs) of hybrid *Cymbidium* Twilight Moon 'Day Light' after culture on a new medium, Teixeira *Cymbidium* (TC) medium (Teixeira da Silva, 2012).

## 8.5. SPICES

Reproductive protocols had been developed by Mathew *et al.* (1999) for synthetic seed production in cultivated *Vanilla planifolia*. Micheli *et al.* (1998) encapsulated apical and nodal buds of olive cv. Moraiolo. Apical buds showed higher viability and re-growth ability than the encapsulated nodal explants. The encapsulated buds could be commercially stored at 4 $^0$C, they observed different behaviour between apical and nodal microcutting after storage; viability and re-growth of nodal microcuttings increased after storage; but was noted opposite trends for apical buds. Micheli *et al.* (2002b) encapsulated somatic embryos of olive cv. Canino and uni-nodal microcuttings of cv. Meraiolo. They reported 56% conversion of somatic embryos into plantlet and it was 90% in case of microcuttings. Excised embryos of mature seeds of cocoa (*Theobroma cacao*) were encapsulated in sodium alginate gel (Sudhakar *et al.*, 2000). Germination of synthetic seeds soon after encapsulation was 97.3%, which declined to 71% and 49% respectively when stored in either wet or dry cotton medium.

## 8.6. FORESTRY AND PLANTATION CROPS

The technique and pattern of the development of somatic embryos in tree species differ, depending on whether the species is ***angiosperm*** or ***gymnosperm***. For the production of artificial seeds in recovery of plant is crucial. In tree species, *Santalum album* (Bapat and Rao, 1988, 1992) and *Pistacia vera* (Onay *et al.*, 1996), the somatic embryos have been encapsulated to produce synthetic seeds. Sprag *et al.* (2002) encapsulated somatic and zygotic embryos of *Pinus patula* in sodium alginate. The alginate beads supplemented with sucrose and charcoal germinated after 14 days and they suggested that the encapsulated embryos could be stored for 20 days. Attree and Fowke (1993) reviewed the advances in synthetic seed technology of conifers. Proliferating embryogenic cultures from conifers consist of immature embryos, which undergo synchronous maturation in the presence of abcisic acid and elevated ***osmoticum***. Improvement in conifer somatic embryo quality has been achieved by identifying the conditions *in vitro*. Conditions of culture were described that yield of mature conifer somatic embryos that posses normal storage proteins and fatty acids and which survive either partial drying or full drying to moisture contents similar to those achieved by mature dehydrated zygotic embryos. Large number of quiescent somatic embryos can be produced throughout the year and stored for germination in the spring, which simplifies the production and provides plants of uniform size.

Ishii *et al.* (1999) encapsulated shoot tips and axillary buds in the production of artificial seeds of *Cedrela odorata, Guazuma crinita* and *Jacaranda mimosifolia*. Artificial seeds were produced by enhancing viable plant material in alginate of *Morus* sp., *Santalum album, Quercus serrata* and *Betula alba* (Kinoshita, 1992). They found that the storage of artificial seeds at 4 $^0$C for upto 40 days had little effect on germination.

Synthetic seeds were prepared by Sudhakar *et al.* (2000) and Sunilkumar *et al.*, (2000) from excised embryos of mature seeds. Storage of those beaded embryos at low temperature of 10 °C and 4 °C retained the viability of synthetic seeds up to four weeks. They concluded that the synthetic seeds produced by this method might be used for exploiting the possibilities of long-term storage of *Hopea parviflora* and *Theobroma cacao* seeds.

Bamboos are generally propagated through vegetative means, but nowadays standard protocols are available for induction of embryogenic callus. Using the technology of somatic embryo induction Mukundakumara and Mathur (1992) prepared synthetic seeds of bamboo (*Dendrocalamus strictus* L.) by emasculating somatic embryos. They could achieve a germination frequency of 96% and 45% in *in vitro* and in soil, respectively. The *in vivo* plantlets conversion frequency was also increased to 56% following an addition of mineral oil on the algenic beads. Germinated plantlets could be raised into plants.

## 8.7. MEDICINAL PLANTS

Propagation and conservation of the pharmaceutically important herbs, shrubs and trees gaining momentum to sustain the medicinal plants biodiversity. Mandal *et al.* (2000, 2001) developed a cost-effective synthetic seed based propagation protocol for basils (*Ocimum* spp.) and to use the synthetic seeds for *in vitro* conservation of the germplasm. They produced synthetic seeds by encapsulating axillary vegetative buds harvested from garden-grown plants of *Ocimum americanum* L. (hoary basil), *O. basilicum* L. (sweet basil), *O. gratissimum* L. (shrubgy basil) and *O. sanctum* L. (sacred basil). They recorded highest frequency shoot emergence and number of shoot per bud were recorded on media containing benzyladenine. Rooted shoots were retrieved from the encapsulated buds of *O. americanum*, *O. basilicum* and *O. sanctum,* whereas shoots of *O. gratissimum* failed to root. They also found that encapsulated explants could be stored for 60 days at 4 $^0$C. Plants retrieved from the encapsulated buds were hardened off and established in soil.

## 8.8. OTHER CROPS

Potential applications of synthetic seed technology for forage crop improvement and propagation have been discussed by many authors (McKersie and Brown, 1996; McKersie *et al*., 1989; Gray *et al*., 1993). McKerrie *et al*. (1989) used somatic embryos to prepare dry synthetic seeds of alfalfa. The embryoids were air dried slowly (over 7 days) or rapidly (over 1 day) to moisture contents of less than 15% and retained fully viable. They reported that the dry somatic embryos could be stored with no loss of viability for 8 months at room temperature and humidity.

Nieves *et al*. (2000) developed synthetic seeds from somatic embryos of sugarcane. Further, Nieves *et al.* (2003a & b) compared the performance of cv. CP 5243 plants derived from artificial seeds with traditional and isolated bud methods. They found that the plants from artificial seeds were taller and had a smaller diameter at 8$^{th}$ month, but those differences disappeared at 12 months. With respect to sugar analysis and yield, no differences in all parameters evaluated were found between artificial seed derived plants and plants derived from the two other methods.

**Table 8.1.** Crop plants in which synthetic seed production and plantlet conversion has been achieved

| Crop | Common name | Propagule* | Reference |
|---|---|---|---|
| **I. Field crops** | | | |
| *Arachis hypogaea* | Ground nut | SE | Padmaja *et al*., 1995 |
| *Brassica capmpestris* | Mustard | SB | Arya *et al*. 1998 |
| *Eleusine coracana* | Finger millet | SE | George and Eapen, 1995 |
| *Hordium vulgare* | Barley | ASE | Datta and Potrykus, 1989 |
| *Oryza sativa* | Rice | SE | Xing *et al*. 1995; Suprsanna *et al*., 1996; Arun Kumar *et al*., 2005; Roy and Mandal, 2008 |

| | | | |
|---|---|---|---|
| Oryza nivara | Wild rice | SE | Jaseela *et al.*, 2009 |
| **II. Vegetable crops** | | | |
| *Asparagus cooperi* | Aspargas | SE | Ghosh and Sen, 1994 |
| *Brassica* spp. | - | SE | Takahata *et al.*, 1992 |
| ***Brassica oleracea* var. *botrytis*** | Cauliflower | SE | Siong, *et al.*, 2012 |
| *Cucumis sativus* | Cucumber | SE | Pellinen *et al.*, 1997 |
| *Daucus carrota* | Carrot | SE | Kitto and Janick, 1982 & 1985; Sakamoto *et al.*,1992; Onishi *et al.*, 1994; Latif *et al.*, 2007 |
| *Ipomea aquatica* | Water Cabbage | AB | Tang *et al.*, 1994a |
| *Ipomea batatus* | Sweet potato | SB | Tang *et al.*, 1994a & b |
| *Picea abies* | Norway spruce | SE | Ghosh and Sen, 1994 |
| *Picea glauca* | Interior spruce | SE | Lulsdorf *et al.*, 1993 |
| *Picea mariana* | Black spruce | SE | Lulsdorf *et al.*, 1993 |
| *Solanum tuberosum* | Potato | SE | Chee and Cantlife, 1992 |
| *Solanum melongena* | Brinjal | SE | Mariani, 1992; Huda and Bari, 2007 |
| **III. Fruits** | | | |
| *Acca sellowiana* | Pineapple guava | SE | Cangahuala-Inocente *et al.*, 2007 |
| *Actinia deliciosa* | Kiwifruit | MC | Andriani *et al*, 2000; Alvarez *et al.* 2002; Piccioni and Standardi, 1995 |
| *Carica papaya* | Papaya | SE | Ye *et al.* 1993 |
| *Citrus reticulata* | Orange | SE | Nieves *et al.*, 1998; Antonietta *et al.*, 1998 |
| *Citrus reticulata* | Orange | NE | Madhav *et al.*, 2002 |
| *Mangifer indica* | Mango | SE | Ara *et al.*, 1999; Litz *et al.*, 1993 |
| *Malus* | Apple | SB | Piccioni and Standardi, 1995; Brischia *et al.*, 2002, Sicurani *et al.*, 2001 |
| *Musa* sp. | Banana | SB | Priya *et al.*, 2003 |
| *Phonix dactylifera* | Date palm | SE | Bekheet *et al.*, 2002, Ibrahim *et al.*, 2003 |
| *Pyrus communis* | Pear | M | Nower *et al.*, 2007 |
| *Rubus idaeus* | Raspberry | SB | Piccioni and Standardi, 1995 |
| *Rubus* sp. | Blackberry | SB | Piccioni and Standardi, 1995 |
| *Vitis vinifera* | Grape | ADB | Hung and Cheng, 1990; Gray and Purohit, 1991 |

| | | | |
|---|---|---|---|
| **V. Ornamental** | | | |
| *Arundina bambusifolia* | - | P | Ahmed and Taludar, 2005 |
| *Coelogyne breviscapa* | Orchid | PLB | Mohanraj *et al.*, 2009 |
| *Cyclamen persicum* | Cyclamen | SE | Ruffoni and Savona, 2006 |
| *Cymbidium gigantieum* | Orchid | PLB | Corrie and Tandon, 1993 |
| *Cymbidium* | Orchid | PLB | Teixeira da Silva, 2012 |
| *Dendrobium candidum* | Orchid | PLB | Zhang *et al.*, 2011 |
| *Dendrobium wardinum* | Orchid | PLB | Sharma *et al.*, 1992 |
| *Eustoma gradiflorum* | - | SE | Ruffoni *et al.*, 1994 |
| *Genista monosperma* | - | SE | Ruffoni *et al.*, 1994, Ruffoni and Savona, 2006 |
| *Geodorum densiflorum* | Orchid | PLB | Datta *et al.*, 1999 |
| *Gypsophila paniculata* | Gypsophila | IDST | Rady and Hanafy, 2004 |
| *Lisianthus russellians* | Lisianthus | SE | Ruffoni and Savona, 2006 |
| *Phalaenopsis hybrida* | Orchid | PLB | Bhattacharjee *et al.*, 1998 |
| *Plumbago indica* | - | AS | Bhattacharryya *et al.*, 2007 |
| *Saintpaulia ionantha* | African violet | L | Daud *et al.*, 2008 |
| **VI. Forest tree and plantation crops** | | | |
| *Batula alba* | - | T/ AB | Kinoshita, 1992 |
| *Dendocalamus strictus* | Bamboo | SE | Mukunthakumaran and Mathur, 1992 |
| *Camellia japonica* | Tea | SE | Janeiro *et al.*, 1997 |
| *Cedrela odorata* | - | ST & AB | Ishii *et al.* 1999 |
| *Eucalyptus citriodora* | Eucalyptus | ADB | Huang and Cheng, 1990 |
| *Guazama crinita* | - | ST & AB | Ishii *et al.* 1999 |
| *Hopea parviflora* | - | ESE | Sunilkumar *et al.*, 2000 |
| *Jacaranda mimiosifolia* | - | ST & AB | Ishii et al. 1999 |
| *Nothofagus alpine* | Rauli-beech | SZE | Cartes *et al.*, 2009 |
| *Pinus patula* | Pine | SE | Sparg *et al.*, 2002 |
| *Pistacia vera* | Pistachio | EEM | Onay *et al.*, 1996 |
| *Quercus serrata* | oak | ST & AB | Kinoshita, 1992 |
| *Quercussuber* | Cork oak | SE | Pintos *et al.*, 2008 |
| *Santalum album* | Sandal wood | SE | Bapat and Rao, 1988; 1990; 1992; Kinoshita, 1992 |
| *Simmondsia chinensis* | Jojoba | SA&AB | Hassan, 2003 |
| *Theobroma cacao* | Cocoa | ESE | Sudhakar *et al.*, 2000 |
| **VII. Spices** | | | |
| *Allium sativum* | Garlic | B | Bekheet SA, 2006 |

| | | | |
|---|---|---|---|
| *Apium graveolence* | Celery | SE | Kim and Janick, 1990; Sakamoto *et al.* 1991; Hayashi, 1993; Onishi *et al.*, 1994; Suchara *et al.*, 1995 |
| Elletaria cardamomum | Cardamom | ST | Ganapathi *et al.*, 1994 |
| *Osmanthus fragrans* | Olive | AB/SE/MC | Micheli *et al.*, 1998 & 2002b |
| *Piper nigrum* | Balck paper | SE | Nair and Gupta, 2007 |
| *Rheobroma cacao* | Cocoa | ESE | Sudhakar *et al.*, 2000 |
| *Vanilla planifolia* | Vanilla | SE | Mathew *et al.*, 1999 |
| *Zhingiber officinale* | Zinger | SB | Sharma *et al.*, 1994 |
| VII. Forage crops | | | |
| *Dactylis glomerata* | Burmuda grass | SE | Gray *et al.*, 1993 |
| *Medicago sativa* | Alfalfa | SE | Redenbaugh *et al.*, 1984; McKersie *et al.*, 1989; Parrott and Bailey, 1993; McKersie and Brown, 1996 |
| *Spartina alterniflora* | Saltmarsh | SE&MP | Utomo *et al.*, 2008 |
| **VIII. Others** | | | |
| Armoracia rusticana | Hors radish | SE | Shigeta and Sato, 1994 |
| *Carum carvi* | - | ST | Furamanova *et al.*, 1991 |
| *Coptis chinensis* | - | SE | Gui *et al.*, 1989 |
| *Dendrobium candidum* | - | SE | Zhang *et al.*, 2001 |
| *Kalanchoe tubiflora* | - | EB | Palmer and Jasrai, 1996 |
| *Murus indica* | Mulbery | AB | Bapat *et al.*, 1987; Bapat and Rao,1990 |
| *Ocimum* spp. | Tulsi | NS | Mandal *et al.*, 2000; 2001 |
| *Pelargonium candidum* | - | - | Zhang *et al.*, 2001 |
| *Saccharum officiarum* | Sugarcane | SE | Nieves *et al.*, 2003a and b |

***SE:** Somatic embryos; **NE:** Nucelar embryo; **ASE:** Androgenic Somatic embryos; **ESE:** Excised seed embryo; **SZE:** Somatic & Zygotic embryos; **EEM:** Embryo & Embryogenic mass; **SB:** Shoot buds; **AB:** Axillary buds; **ADB:** Adventitious buds; **APB:** Apical buds; **EB:** Epiphyllous buds; **ST:** Shoot tips; **SA&AB:** Shoot apical & axillary bud; **MC:** Microcuttings; **M:** Meristem; **PLB:** Protocorm-like bodies; **P:** Protocorm; **MP:** Microplants; **IDST:** *In vitro* derived shoot tips; **L:** Leaf; **NS:** Nodal segments; **B:** Bulblets

## References


Ahmed H, Talukdar MC. 2005. *In vitro* propagation and artificial seed production of *Arundina bambusiflolia*. J Ornamental Hort New-Series. 8(4): 281-283.

Alvarez CR, Gardi T, Standardi A. 2002. Plantlets from encapsulated *in vitro* derived microcuttings of kiwi fruit (cv. Top star R). Agricultura Mediterranea. 132(3-4): 246-252.

Andriani M, Piccioni E, Standardi A. 2000. Effect of different treatments on the conservation of "Hayward" Kiwifruit synthetic seeds to whole plants following encapsulation of *in vitro* derived buds. New Zealand J Crop Hort Sci. 28(1): 59-67.

Antonietta GM, Emanuele P, Alvaro S. 1998. Effects of encapsulation on *Citrus reticulata* Blanco somatic embryo conversion. Plant Cell Tiss Org. Cult. 55(3): 235-237.

Ara H, Jaiswal U, Jaiswal VS. 1999. Germination and plantlet regeneration from encapsulated somatic embryos of mango (*Mangifera indica* L.). Plant Cell Rep. 19: 166-170.

Arun Kumar MB, Vakeswaran V, Krishnasamy V. 2005. Enhancement of synthetic seed conversion to seedlings in hybrid rice. Plant Cell Tiss Org. 81: 97-100.

Arya KR, Beg MU, Kukreja AK. 1998. Microcloning and propagation of endosulfan tolerant genotypes of Mustard brassica campestris through apical shoot bud encapsulation. Indian J Exp Biol. 36: 1161-1164.

Attree SM, Fowke LC. 1993. Embryogeny of gymnosperms: Advances in synthetic seed technology of conifers. Plant Cell Tiss. Org. Cult. 35 (1): 1-35.

Bapat NA, Mahtre M, Rao PS. 1987. Propagation of *Morus* indica L. (Mulberry) by encapsulated shoot buds. Plant Cell Rep. 6: 393-395.

Bapat VA, Rao PS. 1988. Sandalwood plantlets from synthetic seeds. Plant Cell Rep. 7: 434-436.

Bapat VA, Rao PS. 1990. *In vitro* growth of encapsulated axillary buds of mulberry (*Moris indica* L.). Plant Cell Tiss Org Cult. 20: 69-70.

Bapat VA, Rao PS. 1992. Plantlet regeneration from encapsulated and non-encasulated desiccated embryos of a forest tree: Sandalwood (*Santalum album* L.). J Plant Biochem Biotechnol. 1: 109-113.

Bekheet SA, Taha HS, Saker MM, Moursy HA. 2002. A synthetic seed system of date palm through somatic embryos encapsulation. Annals Agret Sci Cairo. 47(1): 325-337.

Bekheet SA. 2006. A synthetic seed method through encapsulation of *in vitro* proliferated bulblets of garlic (*Allium sativum* L.). Arab J Biotech. 9(3): 415-426.

Bhattacharjee S, Khan HA, Reddy PV, Bhattacharjee S. 1998. *In vitro* seed germination, production of synthetic seeds and regeneration of plantlets of *Phalaenopsis* hybrid. Annals Agril Sci, Cairo. 43(2): 539-543.

Bhattacharyya R, Ray A, Gangopadhyay M, Bhattachaya S. 2007. *In vitro* conservation of *Plumbago indica*- a rare medicinal plant. Plant Cell Biotechnol Mol Biol. 8(1/2): 39-46.

Brar DS, Fujimura T, McCouch S, Zapate FJ. 1994. Application of biotechnology in hybrid rice. In: Hybrid Rice Technology: New Development and Future Prospects, Virmani SS (ed.), International Rice Research Institute, Manila, Philippines. pp. 51-62.

Brar DS, Khush GS. 1994. Cell and tissue culture in plant improvement. In: Machanisms in Plant Growth and Improved Productivity, Basra A (ed.), Mercel Dekker Inc., New York, USA. pp. 229-278.

Brischia R, Piccioni E, Standardi A. 2002. Micropropagation and synthetic seed in M.26 application root stock (II): A new protocol for production of encapsulated diffentiating propagules. Cell Tiss Org Cult. 68(2): 137-141.

Cangahuala-Inocente GC, Vesco LL-dal, Steminmacher D, Torrres AC, Guerra MP. 2007. Improvement in somatic embryogenesis protocol in Feijo (*Acca sellowian* (Berg) Burret): introduction, conversion and synthetic seeds.

Cartes PR, Castellanos HB, Rios DL, Sáez KC, Spierccolli SH, Sánchez MO. 2009. Encapsulated somatic and zygotic embryos for for obtaining artificial seeds of rauli-beech (*Nothofagus alpine* (Poepp. & Endli.) Oerst.). Chilean J Agri Res. 69(1): 112-118.

Chee RP, Cantliffe DJ. 1992. Improved production procedures for somatic embryos of sweet potato for a synthetic seed system. Hort Science. 27(12): 1314-1316.

Chetia S, Deka PC, Devi J. 1998. Germination of fresh and stored encapsulated protocorms of orchids. Indian J Exp Biol. 36: 108-111.

Corrie S, Tandon P. 1993. Propagation of *Cymbidium giganteum* Wall. through high frequency conversion of encapsulated protocorms under *in vitro* and *in vivo* conditions. Indian J Exp Biol. 31: 61-64.

Datta KB, Kanjilal B, Sarker D-de, De-Sarkar D. 1999. Artificial seed technology: Development of a protocol in *Geodorum densiflorum* (Lam) Schltr. – an endangered orchid. Current Sci. 76(8): 1142-1145.

Datta SK, Potrykus I. 1989. Artificial seeds in barley: encapsulation of microspore derived embryos. Theor Appl Genet. 77: 820-824.

Daud N, Taha RM, Nasbullah NA. 2008. Artificial seed production from encapsulated micro shoots of *Sainpaulia ionantha* Wendl. (African violet). J Applied Sci. 8: 4662-4667.

Furamanova M, Sowinaska D, Pietrosiuk A. 1991. *Carum carvi* L. (caraway): *In vitro* culture, embryogenesis and the production of aromatic compounds. In: Biotechnology in Agriculture and Forestry, Vol. 5. Medicinal and Arometic Plants III, Bajaj YSP (ed.), Spriger-Verlag, New York, Heidelberg. pp. 176-192.

Ganapathi TR, Bapat VA, Rao PS. 1994. *In vitro* development of encasulted shoot tips of cardamom. Biotech Techniques. 8(4): 239-244.

George L, Epen S. 1995. Encapsulation of somatic embryos of finger millet (*Eleusine coracana* Gaertn.). Indian J Expt Biol. 33: 291-293.

Ghosh B, Sen S. 1994. Plant regeneration from encapsulated embryos of *Asparagus cooperi* Baker. Plant Cell Rep. 13: 381-385.

Gray DJ, Trigiano RN, Conger BV, Redenbaugh K. 1993. Synseeds : Applications of synthetic seeds to crop improvement. pp. 351-366.

Gui YL, Xu TY, Gu SR, Guo ZS, Hou SS, Ke SG, Wu YL, Li HL. 1989. Studies on somatic embryogenesis of *Coptis chinensis.* Acta Botanica Sinica 31(12): 923-927.

Hassan NS. 2003. *In vitro* propagation of Jojoba (*Simmondsia chinensis* L.) through alginate-encapsulated shoot apical and axillary buds. Int J Agri Biol. 5(4): 513-516.

Hayashi M. 1993. Practical application of somatic embryogenesis synthetic seed system. *In*: Current Plant Science and Biotechnology in Agriculture, Vol. 15, You CB, Chen Z, Ding Y (eds.). pp. 305-308.

Huang MJ, Cheng Z. 1990. Preliminary studies on artificial seeds of eucalyptus. In: Studies on Artificial Seeds of Plants, Li X-Q (ed.). Peking Univ Press, Peking, China. pp. 139.

Huda AKMN, Bari MA. 2007. Production of Synthetic Seed by Encapsulating Asexual Embryo in Eggplant (*Solanum melongena* L.). Inter J of Agril Res. 2(9): 832-837.

Ibrahim AI, Gomaa AH, Eldein AS, Faik B, Hamd AM. 2003. Encapsulation of somatic embryos of date palm (*Phonix ductylifera* L.) in synthetic seed coats. Annals Agriil Sci- Moshtohor. 41(1): 299-312.

Ishii K, Maruyama E, Kinoshita I, Jamieson A. 1999. Application of artificial seed technology for regeneration of *in vitro* cultured plantlets of forest trees. In : Proceedings of the Second International Wood Biotechnology Symposium, Canberra, Australia, 10-12 March, 1997, pp. 247-258.

Janeiro LV, Ballester A, Vieitez AM. 1997. In vitro response of encapsulated somatic embryos of Camellia. Plant Cell Tiss. Org Cult. 51: 119-125.

Jaseela F, Sumitha VR, Nair GM. 2009. Somatic embryogenesis and plantlet regeneration in an agronomically important wild rice species *Oryza nivara*. **Asian J Biotechnol.** 1(2): 74-78.

Kim YH, Janick J. 1990. Synthetic seed technology: improving desiccation tolerance of somatic embryos of celery. Acta Horticulturae. 280: 23-28.

Kinoshita I. 1992. The production and use of artificial seed. Res J Food Agri. 15(3): 6-11.

Kitto SK, Janick J. 1982. Polyox as an artificial seed coat for asexual embryos. Horti Sci. 17: 488

Kitto SK, Janick J. 1985. Production of synthetic seeds by encapsulating asexual embryos of carrot. J Amer Hort Sci. 110: 277-282.

Latif Z, Nasir IA, Riazuddin S. 2007. Indigenous production of synthetic seed in *Dacus carota*. Pak J Bot. 39(3): 849-855.

Litz RE, Mathews VH, Moon PH, Pliego-Alfaro F, Yurgalevitch C, Dewald SG. 1993. Somatic embryos of mango (*Mangifera indica* L.). In: Synseeds: applications of synthetic seeds to crop improvement, Redenbaug K (ed.). CRC, Boca Raton. pp. 409-425.

Lulsdorf MM, Tautorus TE, Kikcio SI, Bethune TD, Dunstan DI. 1993. Germination of encapsulated embryos of interior spruce (*Picea glauca engelmannii* Complex) and black spruce (*Picea mariana* Mill). Plant Cell Rep. 12: 385-389.

Madhav MS, Rao BN, Singh S, Deba PC. 2002. Nucellar embryogenesis and artificial seed production in *Citrus reticulata*. Plant Cell Biotechnol Mol Biol. 3(1-2): 77-80.

Malenganba H, Roy BK, Bhattacharya S, Deka PC. 1996. Regulation of encapsulated protocorms of *Phaius tankervilliae* stored at low temperature. Indian J Exp Biol. 34: 802-805.

Mandal J, Patnaik S, Chand PK. 2000. Alginate encapsulation of axillary buds of *Ocimum americunum* L. (hoary basil), *O. basilicum* L. (sweet basil), *O. gratissium* L. (sharubby basil) and *O. sanctum* L. (sacred basil). In vitro Cellular Devl Biol Plant. 36 (4): 487-292.

Mandal J, Patnaik S, Chand PK. 2001. Growth response of the alginate-encapsulated nodal segments of *in vitro* cultured basils. J Herbs Spices Medicinal Plant. 8(4): 65-74.

Mariani P. 1992. Egg plant somatic embryogenesis combined with synthetic seed technology. Capcium Newlet. (special issue), 7-10 September, 1992. pp. 289-294.

Mathew KM, Rao YS, Kumar KP, Madhusoovanan RJ, Potty SN, Kishor PBK. 1999. *In vitro* culture systems in *Vanilla.* In: Plant Tissue Culture and Biotechnology: Emerging Trends. Proceedings of a Symposium held at Hyderabad, India, 29-31 January, 1997. pp. 171-179.

McKersie BD, Brown DCW. 1996. Somatic embryogenesis and artificial seeds in forage legumes. Seed Sci Res. 6(3): 109-126.

McKersie BD, Senaratna T, Bowley SR, Brown DCW, Krochko JE, Bewley JD. 1989. Application of artificial seed technology in the production of hybrid alfalfa (*Medicago sativa* L.). In Vitro Cell Dev Biol. 25: 1183-1188.

Micheli M, Mencuccini M, Standardi A. 1998. Encapsulation of *in vitro* proliferated buds of olive. Adv Hort Sci. 12(4): 163-168.

Micheli M, Standardi A, Dell Orco P, Mencuccini M. 2002b. Preliminary studies on the synthetic seed and encapsulation technologies of *in vitro* derived olive explants. Acta Horticulturae. 586: 911-914.

Mohanraj R, Annanthan R, Bai VN. 2009. Production and storage of synthetic seeds in *Coelogyne breviscapa* Lindi. Asian J Biotechnol. 1(3): 1-5.

Mukunthakumarn S, Mathur J. 1992. Artificial seed production in the male bamboo *Dendrocalamus strictus* L. Plant Sci. 87: 109-113.

Nair RR, Gupta SD. 2007. *In vitro* plant regeneration from encapsulated somatic embryos of black paper (*Piper nigrum* L.). J Plant Sci. 2(3): 283-292.

Nhut DT, Tien TNT, Huong MTN, Hien NTT, Huyen PX, Luan VQ, Teixeira da Silva JA. 2005. Artificial seeds for propagation and preservation of *Cymbidium* spp. Propag Ornam. PLANT 5: 67-73.

Nieves N, Blanco MA, Gonzalez A, Tapia R. 2000. Capacity diffusion of sodium alginate matrix. Effect on encapsulation of sugarcane somatic embryos. Biotechnologia Vegetal. 1: 55-58.

Nieves N, Lorengo JC, Blanco M-de-los A, Gonzalez J, Peralta H, Hernandoz M, Santos R, Concepcion O, Borroto CG, Borroto E, Tapia R, Martinez ME, Funder Z, Gonzalez A, Deloss-A Balanco M. 1998. Artificial endosperm of *Cleopatra tangerine* zygote embryos: a model for somatic embryo encapsulation. Plant Cell Tiss Org Cult. 54 (2): 77-83.

Nieves N, Zambrano Y, Tapia R, Cid M, Pina D, Castillo R. 2003a. Field performance of sugarcane plants obtained from artificial seed in Cuba. Sugarcane International, November, 2003. pp. 23-25.

Nieves N, Zambrano Y, Tapia R, Cid M, Pina D, Castillo R. 2003b. Field performance of artificial seed derived sugarcane plants. Plant Cell Tiss Org Cult. 75(3): 279-282.

Nower AA, Ali EAM, Rizakalla AA. 2007. Synthetic seeds of pear (*Pyrus communis* L.) rootstocks storage *in vitro*. Australian J Basic Appl Sci. 1(3): 262-270.

Onay A, Jeffree CE, Yeoman MM. 1996. Plant regeneration from encapsulated embryoids and an embryogenic mass of pistachio (*Pistaci vera* L.). Plant Cell Rep. 15: 723-726.

Onishi N, Sakamoto Y, Hirosawa T. 1994. Synthetic seeds as an application of mass production of somatic embryos. Plant Cell Tissue Org Cult. 39(2): 137-145.

Padmaja G, Reddy LR, Reddy GM. 1995. Plant regeneration from synthetic seeds of groundnut (*Arachis hypogaea* L.). Indian J Expt Biol. 33: 967-971.

Palmer JP, Jasrai YT. 1996. Precocious growth and effect of ABA: encapsulated buds of *Kalanchoe tubiflora*. J Plant Bioche Biotechnol. 5(2): 103-104.

Parrot WA, Bailey MA. 1993. Characterization of recurrent somatic embryogenesis of alfalfa on auxin-free medium. Plant Cell Tiss Org Cul. 32(1): 69-76.

Pellinen TP, Sorvaris Tahronen R, Sewon P. 1997. Somatic embryogenesis in cucumber (*Cucumis sativus* L.) callus and suspension cultures. Angowandte Botanik. 71(3-4): 116-118.

Piccioni E, Standardi A. 1995. Encapsulation of micropropagated buds of six woody species. Plant Cell Tiss Org Cult. 42(3): 221-226.

Pintos B, Bueno MA, Cuenca B, Manzanera JA. 2008. Synthetic seed production from encapsulated somatic embryos of cork oak (*Quercus suber* L.) and automated growth monitoring. Plant Cell Tiss Organ Cult. 95(2): 217-225.

Priya BT, Arumugam Shakila, Shakila A. 2003. Synthetic seed production in banana : Adv Plant Sci. 16(1): 219-222.

Rady MR, Hanafy MS. 2004. Synthetic seed technology for emasculation and regrowth of *in vitro*-derived *Gypsophila paniculata* L. shoot-tips. Arab J Biotech. 7(2): 251-264.

Redenbaugh K, Nichol J, Kosseler ME, Paasch BD. 1984. Encapsulation of somatic embryos for artificial seed production. *In Vitro* Cell Dev Biol Plant. 20: 256-257.

Roy B, Mandal AB. 2004. Encapsulation of endrogenic embryos and pro-embryos for production of synthetic seeds in elite indica rice var. IR72. In: Abstracts, ISTA Congress Seed Symposium, Budapest, Hungary, May 17-19, 2004. pp. 38.

Roy B, Mandal AB. 2005. Anther culture response in *indica* rice and variation in major agronomic characters among the androclones of a scented cultivar, Karnal local. African J Biotechnol. 4(3): 235-240.

Roy B, Mandal AB. 2008. Development of synthetic seeds involving androgenic embryos and pro-embryos in an elite *indica* rice. Indian J Biotechnol. 7(4): 515-519.

Ruffoni B, Massabo F, Giovannini A. 1994. Artificial seed technology in ornamental species *Lisianthus* and *Genista.* Acta Horticulturae. 362: 277-304.

Ruffoni B, Savona M. 2006. Somatic embryogenesis in floricultural crops: experience of massive propagation of *Lisianthus*, *Genista*, and *Cyclamen*. Floricultre Ornamental Plant Biotechnol. pp. 305-313.

Ruffoni B. 1998. Embryogenic suspension cultures of cyclamen for the production of artificial seeds. Culture Protette. 27(11): 91-96.

Saiprasad GVS, Polisetty R. 2003. Propagation of three orchid genera using encapsulated protocorm-like bodies. *In Vitro* Cell Dev Biol–Plant. 3: 42-48.

Sakamoto Y, Mashiko T, Suzuki A, Kawata H, Iwasaki A, Hayashi M, Kano A, Goto E. 1992. Development of encapsulation technology for synthetic seeds. Acta Horticulturae. 319: 71-76.

Sakamoto Y, Ohnishi N, Hayashi M, Okamoto A, Mashiko T, Sanada M. 1991. Synthetic seeds: the development of a botanical seed analogue. Chemical Regulation of Plants. 26(2): 205-211.

Sharma A, Tandoan P, Kumar. 1992. Regeneration of *D. wardianum* warner (Orchiddaceae) from synthetic seeds. Indian J Exp Biol. 30: 747-748.

Sharma T., Singh BM, Chauhan RS. 1994. Production of disease free encapsulated buds of *Zingiber officinale* Rosc. Plant Cell Rep. 13:300-302.

Shigeta J, Sato K. 1994. Plant regeneration and encapsulation of somatic embryos of horse radish. Plant Science. 102: 109

Sicurani M, Piccioni E, Standardi A. 2001. Micro-propagation preparation of synthetic seed in M.26 apple root stock I: attempts towards saving labour in the production of adventitious shoot tips suitable for encapsulation. Plant Cell Tiss Org Cult. 66(3): 207-216.

Singh F. 1988. Storage of orchid seeds in organic solvents. Gartenbauwissenschaft. 53: 122-124.

Siong PK, Mohajer S, Taha RM. 2012. Production of Artificial seeds derived from encapsulated *in vitro* micro shoots of cauliflower, *Brassica oleracea* var. *botrytis*. Romanian Biotechnological Letters. 17(4): 7549- 7556.

Sparg SG, Jones NB, Staden J-van, Van-Staden J. 2002. Artificial seed from *Pinus patula* somatic embryos. South African J Bot. 68(2): 234-238.

Suchara KI, Kohketsu K, Vogumi N, Kobayashi T. 1995. Efficient production of celery embryos and plantlets released in culture of immobilized gel beads. J Fermentation Bioengineering. 79(6): 585-588.

Sudhkar K, Nagraj BN, Santhoshkumar AV, Sunilkumar KK, Vijayakumar NK. 2000. Studies on the production and storage potential of synthetic seeds in cocoa (*Theobroma cocoa* L.). Seed Res. 28(2): 119-125.

Sunilkumar KK, Sudhakar K, Vijayakumar NK. 2000. An attempt to improve storage life of *Hopea parviflora* seeds through synthetic seed production. Seed Res. 28(2) : 126-130.

Suprasanna P, Ganapathi TR, Rao PS. 1996. Artificial seed in rice (*Oryza sativa* L.): Encapsulation of somatic embryos from mature embryo callus culture. Asia Pacific J Mol Biotechnol. 4(2): 90-93.

Takahata Y, Wakui K, Kaizuma N, Brown DCW. 1992. A dry artificial seed system for *Brassica* crops. Acta Horticulturae. 319: 317-322.

Tang SH, Sun M, Li KP, Zhang QT, Tang PH. 1994b. Studies on artificial seed of sweet potato (*Ipomoea batatus* L. Lam). Acta Agronomica Sinica. 20(6): 746-750.

Tang SH, Sun M, Li KP. 1994a. Studies on artificial seed of *Ipomoea aquatica* Forsk. Acta Horticulturae Sinica. 21(1): 71-75.

Teixeira da Silva JA. 2012. Production of synseed for hybrid *Cymbidium* using protocorm-like bodies. J Fruit Ornamental Plant Res. 20(2): 135-146.

Utomo HS, Wenefrida I, Meche MM, Nash JL. 2008. Synthetic seed as a potential direct delivery system of mass produced somatic embryos in the coastal marsh plant smooth cordgrass (*Spartina alterniflora*). Plant Cell Tiss Organ Cult. 92(3): 281-291.

Vicient CM, Martinez FX. 1998. The potential uses of somatic embryogenesis in agroforestry are not limited to synthetic seed technology. Revista Brasileira de Fisiologia Vegetal. 10(1): 1-12.

Xing XH, Shen YW, Gao MW, Yin DC, Xing XH, Shen VW, Gao MW, Yin DC. 1995. Studies on production of artificial seeds of rice hybrid between indica and japonica. Acta Agronomica Sinica. 21(1): 45-48.

Ye KN, Huang JC, Ji BJ, You CB, Chen ZL, Ding Y. 1993. Hybrid papaya artificial seed production for experimental field. Current Plant Sci Biotechnol Agri. 15: 411-413.

Zhang M, Wei Xiao Y, Huang HR, Zhang M, Wei XY, Huang HR. 2001. A study on the solid enasculation system of the artifical seed of *Dendrobium candidum*. Acta Horticulturae Sinica. 28(5): 435-439.

Zhang YF, Yan S, Zhang Y. 2011. Factors affecting germination and propagators of artificial seeds of *Dendrobium Candidum*. In: International Conference on Agricultural and Biosystems Engineering- Advances in Biomedical Engineering, Vols. 1-2. pp. 404-410.

# 9

# Genetic Stability of Artificial Seed

Artificial seeds have been widely used for micro-propagation of many plant species. Establishment of gene banks for ***ex situ conservation*** of plant germplasm in the form of field gene banks, seed gene banks, *in vitro* collections, and cryogenically preserved tissues is a common practice (Withers 1983; Rao, 2004; Borner, 2006). Alginate encapsulation provides a viable approach for *in vitro* germplasm conservation as it combines the advantages of clonal multiplication with those of seed propagation and storage (Standardi and Piccioni, 1998; Ara *et al*., 2000).

The main aim of artificial seeds is to obtain true to type plants to maintain the ***germplasm***, but, however, during tissue culture, there is a chance of genetic aberration which is commonly known as ***somaclonal variations***. Larkin and Scowcroft (1981) for the first time used the word 'somaclonal variation' to describe these variations as displayed among the plants derived from tissue culture. It is clearly understood that genomes modify themselves when exposed to unfamiliar conditions, and these modifications are best described as programmed loss of cellular control (Phillips *et al*., 1994).

## 9.1. SOMACLONAL VARIATIONS

Somaclonal variation in tissue culture is a common phenomenon which makes it mandatory to check for genetic stability of plants. Somaclonal variations have been defined as genetic and phenotypic variations among clonally propagated plants of a single donor clone and these are manifested as somatically or meiotically stable events. Thus, the plantlets obtained from artificial seeds must be screened for its clonal fidelity. Different factors influencing somaclonal variations, such as, explant source, media composition, the concentration and type of growth regulator in the medium, culture conditions, culture age and genotype of the plant, number and duration of subcultures, effect of stress, genotype and the method of propagation adopted have been reported and studies based on these have been well received (Shenoy and Vasil 1992; Leroy *et al*., 2000; Devarumath *et al*., 2002). The cultural condition itself considered to be mutagenic and plantlets derived from callus, suspension and protoplast often show genotypeic and phynotypic variations (Roy and Mondal; Orbovic *et al*., 2008; Orbovic et al., 2008).

The most commonly observed changes occurring during tissue culture are DNA methylations, chromosome rearrangements and point mutations (Phillips *et al*., 1994). Tissue culture-induced variations have been harnessed to confer desirable traits to cultivars, which include morphological traits, disease resistance, acid tolerance and salt tolerance (Karp, 1992). However, these genetic variations are not sought-after when clonal propagation is the ultimate aim. The potential advantage of synthetic seeds for genetically identical to natural plants was supported by many reports (Nyende *et al*., 2003). The chance of variation is more when plants are regenerated via an intermediate callus phase (Bairu *et al*., 2011) and callus is one of the major propagule for preparation of artificial seeds. The level of synthetic plant growth regulators in the medium is also coupled with somaclonal variation (Martin *et al*., 2006). Plant cell culture results in high frequency of variation in regenerated plants (Larkin and Scowcroft, 1981). Owing to this variation, the resulting plant may not possess the same properties as that of the parent plant.

Somatic variations may result in decline in vigour and regenerability of artificial seeds. In situations where the primary regenerant is the requirement from artificial seeds for direct use, such as, in case of ornamental plants and forest trees, even physiologically induced variations influence the acceptability of the product in the market. Hence, it becomes mandatory to keep a stringent check over the genetic constitution of the micro-propagated plants because it facilitates the screening of undesirable off-types and therefore, nullifies any chances of undesirable accumulation of the variants.

Potential advantages of artificial seeds include their designation as 'genetically identical materials' ease of handling and transportation, along with increased efficiency of *in vitro* propagation in terms of space, time, labour and overall cost. Although artificial seeds have been widely utilized for micro-propagation and conservation of various plant species (Ara *et al*., 2000; Mandal *et al*., 2000 & 2001; Anand and Bansal, 2002; Nyende *et al*., 2003; Manjkhola *et al*., 2005; Singh *et al*., 2006a, b; Narula *et al*., 2007; Faisal and Anis, 2007; Ray and Bhattacharyaa, 2008; Roy and Mandal, 2008; Lata *et al*., 2009), the genetic stability of artificial seed-derived plantlets remains relatively unknown.

The genetic fidelity of plants recovered from cryogenic storage has been assessed in a wide range of plant systems (Haggman *et al*., 1998; Aronen *et al*., 1999; Hirai and Sakai 2000; Scocchi *et al*., 2004; Bekheet *et al*., 2007).

The increased in utilization of artificial seeds for conservation of plant genetic resources and plant propagation thereby necessitates assessment of genetic stability of conserved artificial seeds or plant propagules following their conversion (Dehmer, 2005).

## 9.2. ASSESSMENT OF GENETIC STABILITY

Different approaches have been regularly used for detection of somaclonal variations. These can be detected through morphological, physiological/ biochemical and molecular techniques (Bairu *et al.*, 2011). Of these, molecular techniques have been considered to be superior to morphological and biochemical techniques.

### 9.2.1. DNA-Based Molecular Techniques

Various types of DNA-based molecular markers (RAPD, RFLP, ISSR, AFLP etc.) are available for testing the genetic fidelity of artificial seeds derived plantlets (Sharma *et al.*, 2007; Devarumath *et al.*, 2002).

#### *9.2.1.1. Random Amplified Polymorphic DNA (RAPD)*

RAPD is the most widely used molecular marker for the study of polymorphism and genetic diversity. It is well suited for large-sample-throughout system required for breeding, population genetics and biodiversity. This method is technically simple, quick to perform and requires small amount of DNA and involves no radioactivity. RAPD primers are unable to distinguish between homozygote and heterozygotes. They identify allelic variation among individuals as presence or absence of a particular amplified band. Therefore, RAPDs are considered to be as dominant markers.

In this technique, banding profiles are created using oligonucleotide primers (10 decamer primers) of arbitrary sequence. These primers bind to homologous sequences along the genome. During PCR amplification, these results in production of a sufficient amount of DNA fragments downstream of the primers, and this can be detected on an agarose gel (generally 0.80%) after staining with ethidium bromide.

RAPD markers are simple, fast, cost effective, highly discriminative and reliable. Random amplified polymorphic DNA marker is often used in genetic variation studies in tissue-culture-derived plants when compared to other molecular markers because of the less quantity of DNA required, ease of use, low cost, reliability, less time consuming, and does not require prior knowledge of the nucleotide sequence of the organism under study, with no radioactive probes and no expensive restriction enzymes involved (Williams *et al.* 1990).

RAPD markers have been applied for characterisation of micropropagated forest trees viz. *Picea mariana* (Isabel *et al.*, 1993), *Populus deltoids* (Rani *et al.*, 1995), Oak (Barrett *et al.*, 1997), *Populus tremuloides* (Rahman and Rajora, 2001). Nadha *et al.* (2011) found 15 RAPD primers which produced 84 distinct bands with an average of 5.6 bands per primer. All banding profiles from micropropagated plants were monomorphic and similar to those of the mother plant, thus ascertaining the true nature of the *in vitro* raised plants.

The genetic fidelity of *Rauvolfia* clones raised from synthetic seeds following four weeks of storage at 4 °C were assessed by using RAPD (Faisal *et al.*, 2012). All the RAPD and ISSR profiles from generated plantlets were monomorphic and comparable to the mother plant, which confirms the genetic stability among the clones.

The genetic stability of plantlets derived from encapsulated *Ananas comosus* micro-shoots was proved by random amplified polymorphic DNA (RAPD) and ISSR techniques (Gangopadhyay *et al.*, 2005). Bekheet (2006) reported that in *Allium sativum* both plantlets derived from encapsulated bulblets as well as normally *in vitro* were genetically similar to those that the *in vivo.* RAPD analysis showed the genetic stability of *in vitro* plantlet derived from encapsulated shoot tips of *Dioscorea bulbifera* (Narula *et al.*, 2007). Srivastava *et al.* (2009) reported that *Cineraria maritana*, analysis of the RAPD profiles revealed an average similarity coefficient of 0.944, they confirmed the molecular stability of plants derived from encapsulated microshoots followed by six months of storage. The genetic stability between mother plants and somatic embryo derived synthetic seeds showed resemblance in *Cucumis sativus* and proved by using RAPD markers (Tabassum *et al.*, 2010). In *Picrorhiza kurrooa,* the genetic stability of plants derived from encapsulated microshoots following three months of storage was proved by using cluster analysis of RAPD profile (Mishra *et al.,* 2011a,b,c). Lata *et al.* (2011) reported genetic stability of synthetic seed derived plants of *Cannabis sativa* studied by using ISSR- DAN fingerprinting and gas chromatography (GC) analysis of six major cannabinoids and showed homogeneity in the regrown clones and the mother plant.

DNA amplification of the *in vitro* developed microtillers was recorded in the form of score able banding pattern and compared with parental androgenic plant (Roy and Mandal, 2011). Among the 12-decamer primers used for RAPD analysis, only five could show amplification in the form of discernible bands on the agarose gel. Band sizes in respect of OPD 5, 7, 10, 11 and 20 were found to be between 600 & 1100, 350 & 1707, 400 & 1100, 475 & 1125, 600 & >1717 bp, respectively. Maximum number of five bands was observed in case of OPD-7, whereas minimum two bands were found in respect of OPD-10 (Table 9.1).

**Table 9.1.** Dossier of oligonucleotide decamer primers used

| Primer Designation | Sequence | No. of amplicons |
|---|---|---|
| OPD-5 | 5'TGAGCGGACA3' | 3 |
| OPD-7 | 5'TTGGCACGGG3' | 5 |
| OPD-10 | 5'GGTCTACACC3' | 2 |
| OPD-11 | 5'AGCGCCATTG3' | 3 |
| OPD-20 | 5'ACCCGGTCAC3' | 4 |

The banding pattern for all the five primers in respect of parent and microtillers had no variation (Fig. 9.1). By using arbitrary oligonucleotide primers RAPD generally detects variation in DNA sequences by producing amplicons of varying lengths and thus bear key importance in establishing genetic relationships between parents among the androclones. Using current strategies for evaluating DNA sequence variation independent of copy number changes in repetitive sequences, the present study showed overall genetic uniformity and true-to-type character of the tillers micro-propagated *in vitro* from androgenic plantlets of rice var. IR 72. The monomorphic bands in micro-propagated plants by using different primers were reported earlier (Rout *et al.*, 1998). Monomorphism was also observed by Rani and Raina (1998) in enhanced-axillary-branching culture of mature *Eucalyptus tereticornis* Smith. and *E. camaldulensis* Dehn. plants. Genetic uniformity among the microtillers prospects their utility in genetic transformation work where propagation of transgenic plant especially of androgenic origin would be immensely important (Peng *et al.*, 1992).

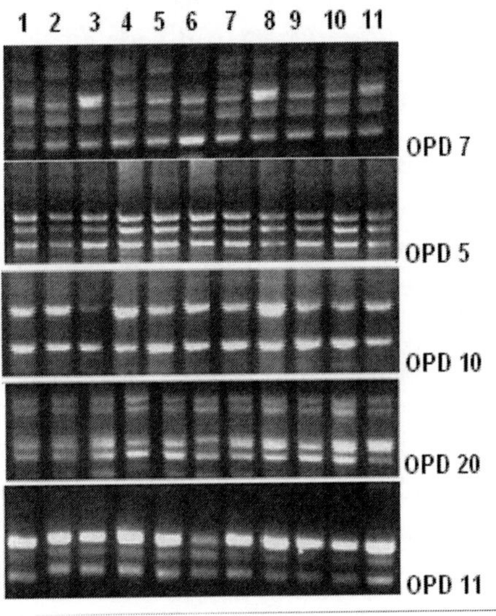


**Fig. 9.1.** Gel electrophoresis of RAPD fragments obtained from microtillers. **Lane 1**: parental clone and **Lanes 2-11**: Randomly selected androgenic microtillers of IR 72.

**OPD-7:** 5′ TTGGCACGGG 3′; **OPD-11:** 5′ AGCGCCATTG 3′; **OPD-5:** 5′ TGAGCGGACA 3′; **OPD-10:** (5′ GGTCTACACC 3′); **OPD-20:** 5′ ACCCGGTCAC 3′.

The genetic stability of micro-propagated clones of ***Gloriosa superba*** L. was evaluated using random amplified polymorphic RAPD analysis (Yadav ***et al.***, 2013). During the study, 50 RAPD primers were screened, out of which 10 RAPD primers produced 49 RAPD clear, distinct and reproducible amplicons. The amplification products of the regenerated plants showed similar banding patterns to that of the mother plant thus demonstrating the homogeneity of the micro-propagated plants. Muthiah ***et al.*** (2013) investigated the efficiency of regrowth of synthetic seeds of ***Bacopa monnieri*** (L.) Pennell after six months of storage along with its clonal fidelity. The genetic stability of regrown plantlets was analysed by using RAPD. A total of 115 bands were produced by 25 RAPD markers with 94% monomorphism. This corresponds to a very low genetic variation pertaining to the clonal nature of the regrown plantlets with that of the mother plant.

***Clonal fidelity*** of the micropropagated plantlets of bamboo has been tested which includes ***Bambusa tulda*** and ***B. balcooa*** using RAPD markers (Das and Pal, 2005; Goyal ***et al.***, 2015). Genetic fidelity of the regenerated plantlets of ***Centella asiatica*** (L.) Urban were confirmed using RAPD analysis, wherein 19 decamer primers produced a total of 60 distinct monomorphic bands (Prasad ***et al.***, 2014). Random amplified polymorphic DNA analysis of tissue culture regenerated plantlets of ***Valeriana officinalis*** L. also indicated no evidence of genetic variation in the tissue culture-raised plants (Ghaderi and Jafari, 2014). RAPD analysis confirmed that the regenerated plants of ***Swertia chirayita*** were genetically identical to their mother plant (Sharma ***et al.***, 2016).

### *9.2.1.2. Inter Simple Sequence Repeat (ISSR) Markers*

Simple sequence repeats are also known as microsatellite. These are ideal DNA markers for genetic mapping and population studies because of their abundance. Inter-simple sequence repeat (ISSR) involves in amplification of DNA segments present at the amplifiable distance in between two identical microsatellite repeat regions oriented in opposite direction. The technique uses microsatellites as primers in a single primer PCR reaction targeting multiple genomic loci to amplify mainly inter-simple sequence repeats of different sizes. The microsatellite repeats used as primers for ISSRs can be di-nucleotide, tri-nucleotide, tetra-nucleotide or penta-nucleotide (A, T, AT, GA, AGG, AAAG etc.).

The primers used can be either unanchored or more usually anchored at 3′ or 5′ end with 1-4 degenerate bases extended into the flanking sequences. The technique is simple, quick and the use of radioactivity is not essential. ISSR use longer perimers (15-30 bp) as compared to RAPD primers, which permits the subsequent use of high annealing temperature leading to higher stringency. ISSR markers usually show high polymorphism. Amenable to detect by both agarose and polyacralimide gel electrophoresis. It exhibits the specificity of microsatellite markers, but need no sequence information for primer synthesis. The primers are not proprietary and can be synthesized by anyone.

ISSR circumvents the requirement for flanking sequence information and thus, has wide applicability in a variety of plants (Srivastava and Gupta, 2008). The SSRs are abundant and scattered throughout the eukaryotic genome (Tautz and Renz, 1984). Hence, ISSR markers are used for detecting clonal fidelity or somaclonal variations that might have resulted during different stages of micropropagation.

The ISSR markers have also been useful in establishing genetic stability of several micropropagated plants, such as cauliflower (Leroy *et al.*, 2000), *Populus tremuloides* (Rahman and Rajora, 2001), *Swertia chirayita* (Joshi and Dhawan, 2007) and *Dictyospermum ovalifolium* (Chandrika *et al.*, 2008), *Ochreinauclea missionis* (Chandrika and Rai, 2009). Negi and Saxena (2010) employed 15 ISSR markers to validate the clonal fidelity of *in vitro* raised *Bambusa balcooa* plantlets through the axillary bud proliferation.

Nadha *et al.* (2011) found 17 ISSR primers, which produced 61 distinct bands in the size range of 300 to 2500 bp. All banding profiles from micropropagated plants were monomorphic and similar to those of the mother plant, thus ascertaining the true nature of the *in vitro* raised plants. The genetic fidelity of *Rauvolfia* clones raised from synthetic seeds following four weeks of storage at 4 °C were assessed by using ISSR markers (Faisal *et al.*, 2012). All the ISSR profiles from generated plantlets were monomorphic and comparable to the mother plant, which confirms the genetic stability among the clones. Muthiah *et al.* (2013) investigated the efficiency of regrowth of *Bacopa monnieri* (L.) Pennell after six months of storage of synthetic seeds along with its clonal fidelity. The genetic stability of regrown plantlets were analysed by using ISSR markers. Twenty ISSR primers produced 130 clear and reproducible amplicons, of which 125 bands were monomorphic. This corresponds to a very low genetic variation pertaining to the clonal nature of the regrown plantlets with that of the mother plant.

The genetic stability of micro-propagated clones of *Gloriosa superba* L. was evaluated using random amplified polymorphic ISSR analysis (Yadav *et al.*, 2013). During the study 30 ISSR primers were screened, out of which 7 ISSR primers produced 49 ISSR clear, distinct and reproducible amplicons. The amplification products of the regenerated plants showed similar banding patterns to that of the mother plant thus demonstrating the homogeneity of the micro-propagated plants, which can be successfully applied for the mass-multiplication and germplasm conservation.

The clonal fidelity among the ***in vitro***-regenerated plantlets of bamboo [***Dendrocalamus strictus*** (Roxb.)] was assessed by Goyal ***et al.*** (2015) using ISSR. The 10 RAPD decamers produced 58 amplicons. All the bands generated were monomorphic. Negi and Saxena (2010) also found genetic stability among the

***in vitro*** regenerated plantlets of ***B. balcooa*** using ISSR. ISSR/RAPD analysis confirmed that the regenerated plants of ***Swertia chirayita*** were genetically identical to their mother plant (Sharma ***et al.***, 2016).

The synthetic seeds were stored in *in vitro* for four months and re-grown under the tissue culture conditions on Murashige and Skoog (MS) medium supplemented with thidiazuron (TDZ 0.2 mg $L^{-1}$). Well rooted plantlets were successfully transferred to a climatic controlled indoor cultivation facility for further cultivation with mother plant. The molecular analysis of genetic fidelity was done using eight ISSR markers. Each tested primer produced clear and scorable amplification products ranging in size from about 176 bp (UBC 817) to 1354 bp (UBC 826) with an average of 5.1 products per primer (Lata *et al.*, 2016). They generated 41 bands, giving rise to monomorphic patterns across all 10 plantlets analyzed. No ISSR polymorphism was observed in the micro-propagated plants.

Lata *et al.* (2011) had assessed the genetic stability of synthetic seeds of *Cannabis sativa* L. during *in vitro* multiplication and storage for six months at different growth conditions using ISSR fingerprinting. Molecular analysis of randomly selected plants from each batch was conducted using 14 ISSR markers. Out of the 14 primers tested, nine produced 40 distinct and reproducible bands. All the ISSR profiles from *in vitro* stored plants were monomorphic and comparable to the mother plant which confirms the genetic stability among the clones.

The generated ISSR profiles from regenerated plantlets of *Withania somnifera* L. with mother plant were monomorphic which confirms the genetic stability among the clones derived from synthetic seeds (Fatima *et al.*, 2013). Thus, the synthetic seed technology could possibly paves the way for the conservation, short-term storage, germplasm exchange with potential storability and limited quarantine restrictions.

### *9.2.1.3. Amplified Fragment Length Polymorphism (AFLP)*

This method combines RFLP and PCR techniques. It has high resolution, which facilitates to identify polymorphism with great accuracy. In amplified fragment length polymorphism (AFLP), genomic DNA is digested with two REs. The choice of REs are very important, one should be frequent cutter and another is the rare cutter. These REs will generate three types of fragments. First one will be fragments which have been cut on both sides of the gene of interest, and these fragments would be highest in number. Second type will be the fragments which have been cut on one side of gene of interest and other cut away from the gene of interest. The third type of fragments, whose cut on away from either side of the gene of interest will be the least among these three types of fragments.

The digested DNA is then amplified using PCR. To make it amplifiable, short adopters for both the restriction enzymes have to be fixed to the fragments. This can be amplified using PCR and analyzed in polyacrylamide gel. The number of polymorphic bands depends on choice of REs and the primer used. The reproducibility of AFLP is ensured by using restriction site specific adapters and adapter specific primers with a variable number of selective nucleotide under stringent amplification conditions. Since, AFLP polymorphism is detected by presence and absence of amplified restriction fragments, it is usually considered dominant markers.

Clonal fidelity among the *in vitro*-regenerated plantlets of bamboo *B. nutans* was obtained by (Mehta *et al.* 2011) employing AFLP. Zarghami *et al.* (2008) evaluated the genetic stability of cryopreserved and non-cryopreserved potato and obtained 97-100% similarity using AFLP markers. Landey *et al.* (2013) observed high genetic and epigenetic stability of *Coffea arabica* plants derived from embryogenic suspensions as well as secondary embryogenesis using AFLP and MSAP markers.

### *9.2.1.4. Sequence Related Amplified Polymorphism (SRAP)*

Recently, a new class of molecular marker known as sequence related amplified polymorphism (SRAP) has been increasingly used for genetic diversity and gene mapping (Li and Quirus 2001). Many researchers had reported the use of SRAP markers in coffee (Mishra *et al.*, 2011a, b, c). The SRAP is a PCR based marker system that preferentially amplifies coding sequences randomly distributed throughout the genome. Moreover SRAP markers possessed multi loci and multi allelic features, which made them more suitable than other marker systems in genetic analysis compared to other marker systems, the multi allelic features of SRAP markers make them more suitable for genetic analysis (Zaefizadeh and Golive, 2009). More recently, SRAP markers were used in assessing the genetic fidelity of some plants obtained through tissue culture (Sun *et al.*, 2014; Li *et al.*, 2014; Peng *et al.*, 2015).

The genetic fidelity of somatic embryogenesis derived plants of *Coffea canephora* (Pierre ex A. Froehner) and the mother plant were tested using sequence related amplified polymorphism (SRAP) markers (Muniswamy *et al.*, 2017). A total of 24 SRAP primers were employed for DNA analysis, which produced a total of 153 clear, distinct and reproducible amplicons of variable size. Cluster analysis revealed more than 95% genetic similarity between the somatic embryogenesis derived plants and the mother plants indicating a high degree of genetic fidelity. Bhatia *et al.* (2011) evaluated genetic fidelity with 100% similarity of *in vitro* propagated plants and mother plant of gerbera using RAPD and ISSR markers.

### *9.2.1.5. Restriction Fragment Length Polymorphisms (RFLP)*

This technique is based on hybridization of DNA molecules where complementary sequence exists. The DNA of different plants is digested with a special class of enzymes called restriction enzymes (RE) or endonuclease. REs usually identify 4, 6, 8 bp long reorganization site. These sites are usually distributed at random throughout genome of an organism. It has the ability to recognize a specific base sequence (restriction site) in the genome. If the requires base sequence is present in the target DNA, the RE will cleave the target DNA at the restriction site. If the number of bases increase in the recognize sequence (restriction site), DNA will be cleaved less frequently. A large piece of DNA will be reduced to a series of smaller fragments of DNA by digestion with a restriction enzyme. The fragments produced by the restriction enzyme will differ in their lengths. Such differences in fragment size, arising out of restriction enzyme digestion of a target DNA, are called, '***Restriction Fragment Length Polymorphism***'. These fragments can be separated according to size subjecting the DNA to agarose gel electrophoresis. The fragments produced will be specific for the target DNA of a plant and restriction enzyme combination. The pattern of fragment sizes of a genotype of a species will differ from that of another genetically different plant genotype of the same species. Thus the RFLP technique can be used to study the extent of genetic similarity as well as dissimilarity among the different genotypes at the molecular level.

Due to large size of nuclear DNA, the restriction enzyme produces a very large number of fragments in a continuous range of sizes. When the digested nuclear DNA is subjected to agarose gel electrophoresis and stained with ethidium bromide, no distinct fragments can be seen in the gel. So, many fragments are produced that the nuclear DNA appears as a continuous smear on the gel. To detect specific fragments by DNA-DNA hybridization, the cloned probe DNA and the DNA of each plant in the gel must be single stranded. Single stranded DNA is produced from double stranded DNA by the process of denaturation using either alkali treatment or high temperature treatment. The single stranded DNA produced is transferred from agarose gel to a nylon membrane by filter or vacuum sucking technique and this process is termed as ***Southern Blotting*** to facilitate hybridization.

Cloned probes are also double stranded. Probes are developed by generating restriction fragments of the species under study. They are made single stranded. Thus under proper temperature and salt concentration, the labeled single stranded probe will hybridize specially to the single stranded DNA on the nitrocellulose membrane which are complementary to each other. Those probes that are hybridized with restriction fragments on the nitrocellulose membrane have sequence homology. Then the nitrocellulose membrane is subjected to autoradiography in order to detect the restriction fragments to which the probe has hybridized.

Application of RFLP markers with appropriately chosen probes gives the possibility to assess the genetic fidelity of micro-propagated plants (Abe *et al.*, 2002; Devarumath *et al.*, 2002). Restriction fragment length polymorphism (RFLP) can be used for screening of genetic stability/unstability in tissue culture raised plants, however, these methods involve the use of expensive enzymes, radioactive labeling and extensive care and hence, appear unsuitable (Kumar *et al.*, 2010). Owing to its high reliability, RFLP marker has been utilized extensively for ascertaining the clonal fidelity of micro-propagated plantlets (Kidwell and Osborn, 1993; Shenoy and Valil, 1995).

#### *9.2.1.6. Microsatellite Markers*

Microsatellite-based marker techniques are sequence targeted. These include simple sequence repeats (SSRs), short tandem repeats (STRs), sequence-tagged microsatellite sites (STMS) and simple sequence length polymorphisms (SSLP). Microsatellites consist of tandemly reiterated short DNA (one to five) sequence motifs which are abundant and occur as interspersed repetitive elements in all eukaryotic as well as in many prokaryote genomes. Microsatellite marker techniques use the intra- and inter-individual variation in microsatellites or simple sequence repeat regions for fingerprinting analyses (Agarwal *et al.*, 2008).

Microsatellite markers have been found immensely useful in establishing genetic stability of several micropropagated plants such as sorghum (Zhang *et al.*, 2010), trembling aspen (Rahman and Rajora, 2001), rice (Gao *et al.*, 2009), bananas (Hautea *et al.*, 2004; Ray *et al.*, 2006), grapevine (Welter *et al.*, 2007), wheat (Khlestkina *et al.*, 2010) and sugarcane (Singh *et al.*, 2008). In comparison to AFLP and RFLP, the microsatellite marker technique such as ISSR is cost efficient, overcomes hazards of radioactivity and requires lesser amounts of DNA (Zietkiewicz *et al.*, 1994). In addition, these markers are highly reproducible and becoming more popular due to their co-dominant inheritance nature, high abundance in organisms, enormous extent of allelic diversity as well as the ease of assessing microsatellite size variation using PCR with pairs of flanking primers (Li *et al.*, 2002; Weising *et al.*, 2005; Agarwal *et al.*, 2008). However, a major drawback for the use of microsatellites is that the development of primers is time-consuming (Squirrell *et al.*, 2003).

### 9.2.2. Biochemical Analysis

Any biochemical compounds (hormone, enzyme, antibody, or other biochemical compounds) that is sufficiently altered during plant tissue culture can be used as marker to find out the genetic similarities or dissimilarities among the tissue cultured regenerated or synthetic seed derived plants. Relative to morphological detection, the use of physiological responses and/or biochemical tests for detecting variants is faster and can be carried out at juvenile stages to lower the possible economic loss (Israeli *et al.*, 1995). The response of plants to physiological factors such as hormones and light can be used as basis to differentiate between normal and variant somaclones (Peyvandi *et al.*, 2009).

Many authors have used different biochemical tests to distinguish amongst somaclones. This approach has been very useful to quantify some interesting somaclonal variants (Daub and Jenns, 1989; Kole and Chawla, 1993; Thomas *et al.*, 2006). However, the application of most biochemical tests is complex and requires high expertise. Beside, most of the biochemical tests are done on small scale in the laboratory using in vitro techniques. Most favourable results obtained have not been successfully implemented for commercial purposes (Daub, 1986).

#### *9.2.2.1. Isozyme Analysis*

Among the biochemical compounds, isozymes were widely used to detect the genetic fidelity of synthetic seed derived plants. Isozyme analysis is a powerful biochemical technique with numerous applications in plant science. It has long been used by geneticists to study the population genetics of higher plants. More recently this procedure has been adopted and is now being used routinely to settle taxonomic disputes, identify 'unknown' cultures, 'fingerprint' plant cultivars, analyze genetic variability or similarity etc.

***Isozymes*** are defined as multiple molecular forms of a single enzyme. These forms usually have similar, if not identical, enzymatic properties, but slightly different amino acid compositions due to differences in the nucleotide sequence of the DNA that codes for the protein. Often the only difference among isozymes is the substitution of one to several amino acids. Only those isozymes that have large variations in size or shape or that which differs in net charge can be separated by electrophoresis. Differences in net charge can occur when a basic amino acid is substituted for an acidic amino acid.

Use of isozyme markers was widely used in the past for cultivar identification in plants. Isozymes provide potentially powerful and reliable tools in resolving genetically related divergence and have been used as molecular markers in genetic, phylogenetic, taxonomic and evolutionary studies (Moss, 1982; Richardson *et al.*, 1986).

Only 28.7% of all amino acid substitutions will change the net charge of a protein. Some amino acid substitutions that do not involve charge differences can also affect the electrophoretic mobility of a protein, presumably by altering the tertiary structure of the enzyme.4 Thus, about one third of all single amino acid substitutions will be electrophoretically detectable, and several simultaneous substitutions can cancel out the effect. Isozyme analysis therefore provides a very conservative estimate of the extent of genetic variability within a population. Detectable isozymes can arise from three different genetic and biochemical conditions: (1) multiple alleles at a single locus, (2) single or multiple alleles at multiple loci, and (3) secondary isozymes, usually arising from post-translational processing.

In tea, isozymes have been analysed by several workers (Hairong *et al.*, 1987; Xu *et al.*, 1987; Anderson, 1994; Singh and Ravindranath, 1994). Among the isozymes, peroxidase and esterase have been studied in several tea cultivars. Some of the other isozymes like tetrazolium oxidase, aspartate aminotransferase and alpha-amylase were also studied among 7 different tea cultivars along with 3 different species (Mondal *et al.*, 2004). The banding pattern revealed both quantitative and qualitative variation amongst the different species and their clones; moreover, between these enzymes, tetrazolium oxidase exhibited highest variability. Although isozyme analyses have been used as a marker, in general, isozyme studies in tea were restricted to a few enzymes with inadequate polymorphism (Wachira *et al.*, 1995), and thus with the advancement of newer techniques of molecular biology such efforts were shifted to DNA based markers.

To check the clonal fidelity of micro-propagated plantlets derived through anther culture, a few isozymes were analyzed by Mandal and Roy (2003); Roy and Mandal (2005) to check the genetic stability of the androclones. Peroxidase profile showed no change in the banding pattern in all the samples. All of them showed four bands with Rf 0.12, 0.28, 0.41 and 0.68, respectively (Fig. 9.2). Malate dehydrogenase displayed six bands with Rf values 0.03, 0.15, 0.42, 0.54, 0.64 and 0.69, respectively and all the loci were found to be monomorphic in parent androclones and microtillers of the sister clones.

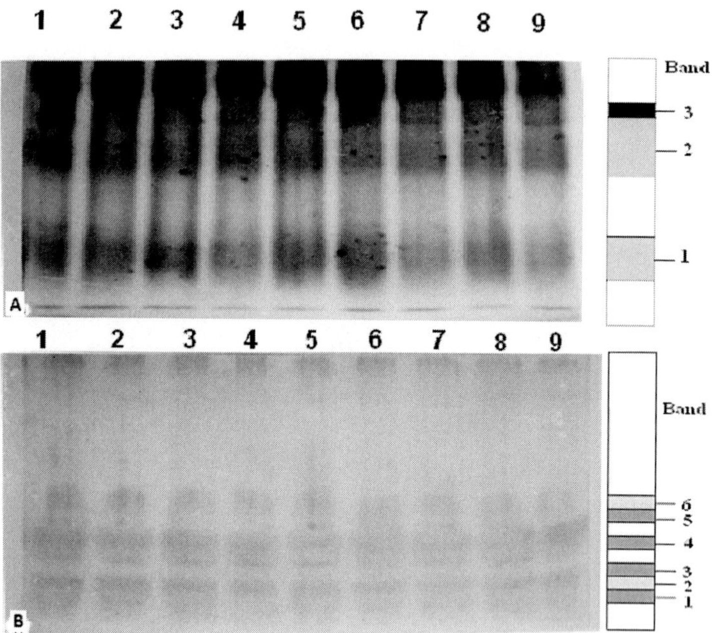


**Fig. 9.2.** Profiling of microtillers in respect of isozymes. **A)** Peroxidase: Lane 1- Parent androclone; Lanes 2-9: Randomly selected microtillers; **B)** Malate dehydrogenase: Lane 1- Parent androclone; Lanes 2-9: Randomly selected microtillers.

Specific variations in metabolic status generally results in discernible differences in isozyme pattern. Staining was performed individually for four isozymes viz., peroxidase and malate dehydrogenenase. Similar banding pattern was observed in all the samples in respect of individual isozymes. This confirms the true to type nature of the micropropagated plantlets derived from IR 72.

### *9.2.2.2. Bioactive Centellosides*

The HPLC analysis of leaf extracts of synthetic seed-derived plant progeny and the mother plant of *Centella asiatica* (L.) Urban showed qualitative and quantitative uniformity in terms of four major bioactive centellosides namely, asiaticoside, madecassoside, asiatic acid, and madecassic acid (Prasad *et al.*, 2014). By means of HPLC biochemical analysis, bacoside-A content in the regrown plants from synthetic seed of *Bacopa monnieri* was found to be comparable with that of the mother plant (Muthiah *et al.*, 2013). As so far, bacoside-A is considered to be the most important secondary metabolite of the plant used commercially in large quantities. HPLC analysis revealed the homogeneity in bacoside profiles and contents in mother and regrown plants.

The quantity of the four different components of bacoside A (bacoside A3, bacopaside II, jujubogenin isomer of bacopasaponin C, and bacopasaponin C) in the mother and regrown plants is shown were similar. Quantification of phytochemicals in plants regrown after low temperature is inevitable so that it is confirmed that the storage process has maintained the chemical stability of the plant. However, only a very few reports are available in this regard. Lata *et al.* (2011) estimated the cannabinoids content of a *Cannabis sativa* mother plant and randomly selected clones propagated through synthetic seeds following storage for 6 months. They showed that the level of Δ9 -tetrahydrocannabinol, the major psychoactive compound, in the mature buds of the synthetic seeds' raised clones of *C. sativa* plants was comparable to that of the mother plant.

### *9.2.2.3. Gibberellic Acid*

The response of plants to physiological factors such as hormones can be used as basis to differentiate between normal and variant somaclones. Gibberellic acid regulates growth and influences various developmental processes such as stem elongation and enzyme induction. Therefore, disturbances in gibberellic acid metabolism and level have been suggested as one of the possible indicators of clonal fidelity or somaclonal variation in higher plants (Phinney 1985; Sandoval *et al.*, 1995). Based on responses to exogenous gibberellic acid, dwarfs of many plant species have been classified as either gibberellic acid-responsive or gibberellic acid-non responsive (Graebe, 2003). For example, Damasco *et al.* (1996) observed that normal banana plants showed significantly greater leaf-sheath elongation in response to gibberellic acid treatment than dwarfs and elongation in normal banana plants was two-fold greater than the dwarfs. Similarly, Sandoval *et al.* (1995) reported a hormonal analysis based on endogenous gibberellins which could be useful to understand and characterize the inter-varietal differences linked to height in *Musa* spp. However, this technique is mostly useful and has been proved to be very effective for plantlets at the de-flasking stage (Damasco *et al.*, 1996).

### *9.2.2.4. Photoinhibition*

Photoinhibition is a state of physiological stress and is expressed as a decline in photosynthetic capability of oxygen evolving photosynthetic organisms due to excessive illumination and plays an important role in plants (Adir *et al.*, 2003). Generally, shade tolerant plants are more susceptible to photoinhibition than sun-loving plants (Long *et al.*, 1994). Senevirathna *et al.* (2008) suggested a mild shade environment in which the level of photosynthetic photon flux density is high enough to saturate carbondioxide assimilation and low enough to induce shade acclimation will help to optimize the photosynthetic productivity in plants. Based on photoinhibition, Damasco *et al.* (1997) evaluated the responses of tissue cultured normal and dwarf off-type Cavendish bananas to suboptimal temperatures under field and controlled environmental conditions. The dwarf off-types showed improved tolerance to low temperature and light compared to the normal somaclones.

#### *9.2.2.5. Pigment Synthesis*

The synthesis of pigment such as chlorophyll, carotenoids, anthocyanins can be used as basis for detecting somaclonal variants (Shah *et al*., 2003). For example, Mujib (2005) observed that pineapple variants showed significantly lower chlorophyll than normal regenerants. Similarly, the total carotenoid content varied greatly between normal and variant somaclones (Wang *et al*., 2007).

### 9.2.3. Morphological Characterization

Morphological characters have long been used to identify species, genera, and families in plants. To find out the similarity or genetic fidelity among the synthetic seed derived plants may be done through morphological characterization by visual observation of different plant traits, such as plant height, branching or tillering pattern, days/month/year to flowering, leaf (size, shape, pattern, colour etc.), flower (size, shape, color, pattern etc.), fruits/ grains (size, shape, clour, test eight etc.), inflorescence, potential yield etc. Morphological characterization is generally performed following standard 'descriptor' available for particular crop or species. It was also found that morphological traits often controlled by environment and environment × genomic interaction. Therefore, the changes need to be investigated more thoroughly and for at least two progenies in *ex vitro* condition (Jarret and Gawel, 1995; Mandal *et al*., 2001).

Morphological characteristics which were similar in both seed-propagated and micro-propagated plants were observed by Mallaya and Ravishankar (2013) in *in vitro* regenerated plantlets of egg plants (*Solanum melongena* L., cv. Arka Shirish). Synthetic seeds of *Centella asiatica* (L.) Urban were kept on moist filter paper in sealed petri plates and stored at 25 ± 3 °C temperature for 200 days (Prasad *et al*., 2014). The synthetic seeds thus conserved recorded more than 85% germination and plantlet conversion frequency upon shifting onto a hormone-free Murashige and Skoog medium. The rooted plantlets recorded 85-90% establishment in soil and did not show any morphological variation when compared with mother plants.

In Brazil, an experimental field was established to evaluate the tissue cultured plants of Bourbon obtained from somatic embryos produced in both solid and liquid media and compare their performance with the seedlings of the same line (Sondahl *et al.*, 1999). In another study, five robusta clones are being field tested in five coffee producing countries (4000 plants/location): Philippines, Thailand, Mexico, Nigeria and Brazil. The visual inspection of 8000 plants under field conditions in Philippines revealed that all the micro-propagated *robusta* plants have normal vegetative and reproductive growth with normal flowering and fruit set after 2 years of field planting (Sondahl *et al.*, 2001). Similarly, Etienne *et al.* (1999) and Etienne and Bertrand (2001) evaluated 20000 field grown *arabica* tissue culture plants in four Central American countries and reported the trueness-to-type and agronomic characteristics of tissue cultured plants produced by embryogenic cell suspensions. Muniswamy *et al.* (2002) reported the field performance of tissue cultured plants of *Coffea arabica.*

Jain and Datta (1992) compared various morphological characters such as leaf shape, petiole length, area of leaf lamina and internodal distance in both *in vitro* raised plants of *Morus bombycis* cultivar, Shimanochi and the vegetative propagated saplings and observed that there was no significant quantitative variation between the characters.

However, morphological traits are often strongly influenced by environmental factors and may not reflect the true genetic composition of a plant (Mandal *et al.*, 2001). In addition, morphological markers used for phenotypic characters are limited in number, often developmentally regulated and easily affected by environmental factors (Cloutier and Landry, 1994). Furthermore, the detection of variants using morphological features is often mostly feasible for fully established plants either in the field or greenhouse. This is not an ideal technique for commercial application due to cost implications (Israeli *et al.*, 1995).

## References


Abe T, Li N, Togashi A, Sasahara T. 2002. Large deletions in chloroplast DNA of rice calli after long-term culture. J Plant Physiol. 159: 917-923.

Agarwal M, Shrivastava N, Padh H. 2008. Advances in molecular marker techniques and their applications in plant sciences. Plant Cell Rep. 27: 617-631.

Anand Y, Bansal YK. 2002. Synthetic seeds: a novel approach of in vitro plantlet formation in Vasaka (*Adhatoda vasica* Nees.). Plant Biotech. 19:159-162

Anderson S. 1994. Isozyme analysis to differentiate between tea clones. In: lightings bulletin Institut vir Tropiese en Subtropiese Gewasse. p. 15.

Ara H, Jaiswal U, Jaiswal VS. 2000. Synthetic seed: prospects and limitations. Curr Sci. 78: 1438-1444.

Aronen TS, Krajnakova J, Ha¨ggman HM, Ryyna¨nen LA. 1999. Genetic fidelity of cryopreserved embryogenic cultures of openpollinated Abies cephalonica. Plant Sci. 142: 163-172.

Bairu MW, Aremu AO, Staden JV. 2011. Somaclonal variation in plants: causes and detection methods. Plant Growth Regul. 63: 147-173.

Barrett C, Lefort F, Douglas GC. 1997. Genetic characterization of oak seedlings, epicormic, crown and micropropagated shoots from mature trees by RAPD and microsatellite PCR. Scientia Horti. 70: 319-330.

Bekheet SA, Taha HS, Saker MM, Solliman ME. 2007. Application of cryopreservation technique for In vitro grown Date Palm (*Phoenix dactylifera* L.) cultures. J Appl Sci Res. 3: 859-866.

Bekheet SA. 2006. A synthetic seed method through encapsulation of *in vitro* proliferated bulblets of garlic (*Allium sativum* L.) Arab J Biotech. 9(3): 415-426.

Bhatia R, Singh KP, Sharma TR, Jhang T. 2011. Evaluation of the genetic fidelity of in vitro-propagated gerbera (*Gerbera jamesonii* Bolus) using DNA-based markers. Plant Cell Tissue Organ Cult. 104: 131-135.

Borner A. 2006. Preservation of plant genetic resources in the biotechnology era. Biotechnol J. 1: 1393-1404.

Chandrika M, Rai VR. 2009. Genetic fidelity in micropropagated plantlets of *Ochreinauclea missionis* an endemic, threatened and medicinal tree using ISSR markers. African J. Biotechnol. 8(13): 2933-2938.

Chandrika M, Thoyajaksha, Rai VR, Ramachandra K. 2008. Assessment of genetic stability of *in vitro* grown *Dictyospermum ovalifolium*. Biol Plant. 52(4): 735-739.

Chandrika M, Thoyajaksha, Ravishankar rai V, Ramachandra kini K. 2008. Assessment of genetic stability of *in vitro* grown *Dictyospermum ovalifolium*. Biol Plant. 52(4): 735-739.

Cloutier S, Landry B. 1994. Molecular markers applied to plant tissue culture. In Vitro Cell Dev Biol Plant. 30: 32-39.

Damasco OP, Godwin ID, Smith MK, Adkins SW. 1996. Gibberellic acid detection of dwarf off-types in micropropagated Cavendish bananas. Aust J Exp Agric. 36: 237-341.

Das M, Pal A. 2005. Clonal propagation and production of genetically uniform regenerants from axillary meristems of adult bamboo. J Plant Biochem Biot. 14(2): 185-188.

Daub ME, Jenns AE. 1989. Field and greenhouse analysis of variation for disease resistance in tobacco somaclones. Phytopathology. 79: 600-605.

Daub ME. 1986. Tissue culture and the selection of resistance to pathogens. Annu Rev Phytopathol. 24: 159-186.

Dehmer K. 2005. Identification of genetic diversity as a precondition of efficient preservation in genebanks. Schr Genet Ressour. 24: 1-6.

Devarumath R, Nandy S, Rani V, Marimuthu S, Muraleedharan N, Raina S. 2002. RAPD, ISSR and AFLP fingerprint as useful markers to evaluate genetic integrity of micropropagated plants of three diploid and triploid elite tea clones representing *Camellia sinensis* (China type) and *C. assamica* ssp. *Assamica* (Assam-India type). Plant Cell Rep. 21: 166-173.

Etienne H, Barry-Etienne D, Vasquez N, Berthouly M. 1999. Aportes de la cionbiotechynological mejoramiento genetico Del cafe el ejemplo de la multiplication por embrigenesis somatica de hybridos F1 en America Central. In: B. Bertrand and B. Rapidel. Editors. IICA, Sanjose Desafios de Cafecultura en centroamerica. pp. 457-493.

Etienne H, Bertrand B. 2001. Trueness-to-type and agronomic characters of Coffea arabicacell suspension technique. Tree Physiol. 21: 1031-1038.

Faisal M, Abdulrahman A. Alatar, Ahmad N, Anis M, Hegazy AK. 2012. Assessment of Genetic Fidelity in *Rauvolfia serpentine* Plantlets Grown from Synthetic (Encapsulated) Seeds Following *in Vitro* Storage at 4 °C. Molecules. 17(5): 5050-5061.

Faisal M, Anis M. 2007. Regeneration of plants from alginateencapsulated shoots of *Tylophora indica* (Burm.f.) Merrill, an endangered medicinal plant. J Hort Sci Biotechnol. 82: 351-354.

Fatimaa N, Ahmada N, Anis M, Ahmad I. 2013. An improved in vitro encapsulation protocol, biochemical analysis and genetic integrity using DNA based molecular markers in regenerated plants of *Withania somnifera* L. Industrial Crops and Products 50: 468-477.

Gangopadhyay G, Bandyopadhyay T, Poddar R, Gangopadhyay SB, Mukherjee KK. 2005. Encapsulation of pineapple micro shoots in alginate beads for temporary storage. Curr Sci. 88: 972-977.

Gao D-Y, Vallejo V, He B, Gai Y-C, Sun L-H. 2009. Detection of DNA changes in somaclonal mutants of rice using SSR markers and transposon display. Plant Cell Tissue Organ Cult. 98: 187-196.

Ghaderi N, Jafari M. 2014. Efficient plant regeneration, genetic fidelity and high-level accumulation of two pharmaceutical compounds in regenerated plants of *Valeriana officinalis* L. South African J Bot. 92: 19-27.

Goyal AK, Pradhan S, Basistha BC, Sen A. 2015. Micropropagation and assessment of genetic fidelity of *Dendrocalamus strictus* (Roxb.) nees using RAPD and ISSR markers. Biotech. 5: 473-482.

Graebe JE. 2003. Gibberellin biosynthesis and control. Annu Rev Plant Physiol. 38: 419-65.

Haggman HM, Ryyna¨nen LA, Aronen TS, Krajnakova J. 1998. Cryopreservation of embryogenic cultures of Scots pine. Plant Cell Tissue Organ Cult. 54: 45-53.

Hairong X, Qiqing T, Wanfanz Z. 1987. Studies on the genetic tendency of tea plant hybrid generation using isozyme technique. In: Proc. Int. Symp Tea Quality and Human Health, Hangzhou, China. pp. 21-25.

Hautea DM, Molina GC, Balatero CH, Coronado NB, Perez EB, Alvarez MTH, Canama AO, Akuba RH, Quilloy RB, Frankie RB, Caspillo CS. 2004. Analysis of induced mutants of Philippine bananas with molecular markers. In: Jain SM, Swennen R (eds.) Banana improvement: cellular, molecular biology, and induced mutations. Science Publishers, Inc., Enfield. pp. 45-58.

Hirai D, Sakai A. 2000. Cryopreservation of in vitro-grown meristems of potato (S*olanum tuberosum* L.) by encapsulationvitrification. In: Engelmann F, Takagi H (eds.) Cryopreservation of tropical plant germplasm. Current research progress and application, JIRCAS-IPGRI, Rome. pp. 205-211.

Isabel NL, Tremblay M, Michaud F, Tremblay M, Bousquet J. 1993. RAPDs as an aid to evaluate the genetic integrity of somatic embryogenesis derived population of *Picea mariana* (Mill.) B.S.P. Theor Appl Genet. 86: 81-87.

Israeli Y, Ben-Bassat D, Reuveni O. 1996. Selection of stable bananaclones which do not produce dwarf somaclonal variants during in vitro culture. Sci Hortic. 67: 197-205.

Israeli Y, Lahav E, Reuveni O. 1995. *In vitro* culture of bananas. In: Gowen S (ed.) Bananas and plantians. Chapman and Hall, London. pp. 147-178.

Jain AK, Datta RK. 1992. Shoot organogenesis and plant regeneration in mulberry (Morusbombycis Koidz): factors influencing morphogenetic potential in callus cultures. Plant Cell Tiss Organ Cult. 29: 43-50.

Jarret RL, Gawel N. 1995. Molecular Markers, Genetic Diversity and Systematics in Musa. In: Bananas and Plantains, Gowen S. (Ed.). Chapman and Hall, London, UK. pp: 66-83.

Joshi P, Dhawan V. 2007. Assessment of genetic fidelity of micropropagated *Swertia chirayita* plantlets by ISSR marker assay. Biol Plant. 51(1): 22-26.

Karp A. 1992. The role of growth regulators in somaclonal variation. Br Soc Plant Growth Regul Annu Bull. 2: 1-9.

Khlestkina E, Ro¨der M, Pshenichnikova T, Bo¨rner A. 2010. Functional diversity at the Rc (red coleoptile) gene in bread wheat. Mol Breeding. 25: 125-132.

Kidwell KK, Osborn TC. 1993. Variation among alfalfa somaclones in copy number of repeated DNA sequences. Genome. 36: 906-912.

Kole P, Chawla H. 1993. Variation of *Helminthosporium* resistance and biochemical and cytological characteristics in somaclonal generations of barley. Biol Plant. 35: 81-86.

Kumar N, Modi AR, Singh AS, Gajera BB, Patel AR, Mukesh P. 2010. Patel and Naraynan Subhash. 2010. Assessment of genetic fidelity of micropropagated date palm (*Phoenix dactylifera* L.) plants by RAPD and ISSR markers assay. Physiol Mol Biol Plants. 16(2): 207-2013.

Landey Bobadilla R, Cenci A, Georget F, Bertrand B, Camayo G, Dechamp E, Herrera JC, Santoni S, Lashermes P, Simpson J, Etienne H. 2013. High Genetic and Epigenetic Stability in Coffea arabica plants derived from embryogenic suspensions and secondary embryogenesis as revealed by AFLP, MSAP and the phenotypic variation rate, PLoS One, 8 (2), e56372. doi:10.1371/journal.pone.0056372.

Larkin PJ, Scowcroft WR. 1981. Somaclonal variation- a novel source of variability from cell cultures for plant improvement. Theor Appl Genet. 60: 197-214.

Lata H, Chandra S, Khan IA, ElSohly MA. 2009. Propagation through alginate encapsulation of axillary buds of *Cannabis sativa* L.- an important medicinal plant. Physiol Mol Biol Plants. 15: 79-86.

Lata H, Chandra S, Natascha T, Khan IA, ElSohly MA. 2011. Molecular analysis of genetic fidelity in *Cannabis sativa* L. plants grown from synthetic (encapsulated) seeds following *in vitro* storage. Biotechnol Lett. 33: 2503-2508.

Lata M, Chandra S, Techen N, Wanf YA, Elsohli MA, Khan IA. 2016. Genetic fidelity of *Stevia rebaudiana* Bertoni plants grown from synthetic seeds following *in vitro* storage. Planta Medica. 82(05).

Leroy XJ, Leon K, Charles G, Branchard M. 2000. Cauliflower somatic embryogenesis and analysis of regenerant stability by ISSRs. Plant Cell Rep. 19(11): 1102-1107.

Li G, Quirus CF. 2001. Sequence–related amplified polymorphism (SRAP), a new marker system based on a simple PCR reaction: Its application to mapping and gene tagging in Brassica, Theor Appl Genet. 103: 455-461.

Li X, Wang X, Luan C, Yang J, Cheng S, Dai Z, Mei P, Huang C. 2014. Somatic embryogenesis from mature zygotic embryos of *Distylium chinense* (Fr.) Diels and assessment of genetic fidelity of regenerated plants by SRAP markers. Plant Growth Regul. 74: 11-21.

Li Y-C, Korol AB, Fahima T, Beiles A, Nevo E. 2002. Microsatellites: genomic distribution, putative functions and mutational mechanisms: a review. Mol Ecol. 11: 2453-2465.

Mandal A, Maiti A, Chowdhury B, Elanchezhian R. 2001. Isoenzyme markers in varietal identification of banana. *In Vitro* Cell Dev Biol Plant. 37: 599-604.

Mandal AB, Roy B. 2003. *In vitro* selection of callus tolerant to high levels of Fe-toxicity in rice and their isozyme profiling. J Genet Breed. 57: 325-340.

Mandal J, Patnaik S, Chand PK. 2000. Alginate encapsulation of axillary buds of *Ocimum americunum* L. (hoary basil), *O. basilicum* L. (sweet basil), *O. gratissium* L. (sharubby basil) and *O. sanctum* L. (sacred basil). In vitro Cellular Devl Biol Plant. 36 (4): 487-292.

Mandal J, Patnaik S, Chand PK. 2001. Growth response of the alginate-encapsulated nodal segments of *in vitro* cultured basils. J Herbs Spices Medicinal Plant. 8(4): 65-74.

Mandal J, Pattnaik S, Chand PK. 2000. Alginate encapsulation of axillary buds of *Ocimum americanum* L. (Hoary Basil), *O. basilicum* L. (Sweet Basil), *O. gratissimum* L. (Shrubby Basil), and *O. sanctum* L. (Sacred Basil). In Vitro Cell Dev Biol Plant. 36: 287-292.

Manjkhola S, Uppeandra D, Meena J. 2005. Organogenesis, embryogenesis and synthetic seed production in *Arnebia euchroma*- a critically endangered medicinal plant of the Himalaya. In vitro Cell Dev Biol Plant. 41: 244-248

Martin K, Pachathundikandi S, Zhang C, Slater A, Madassery J. 2006. RAPD analysis of a variant of banana (*Musa* sp.) cv. grande naine and its propagation via shoot tip culture. *In Vitro* Cell Dev Biol Plant. 42: 188-192.

Mehta R, Sharma V, Sood A, Sharma M, Sharma RK. 2011. Induction of somatic embryogenesis and analysis of genetic fidelity of in vitro-derived plantlets of *Bambusa nutans* Wall., using AFLP markers. Eur J For Res. 130(5):729-736.

Mishra MK, Bhat AM, Suresh N, Satheesh Kumar S, Padmajyothi D, Surya Prakash N, Kumar A, Jaayarama. 2011c. Molecular genetic analysis of arabica coffee hybrids using SRAP marker approach. J Plantation Crops. 39(1): 41-47.

Mishra MK, Nishani S, Jayarama. 2011a. Molecular identification and genetic relationships among coffee species (*Coffea* L.) inferred from ISSR and SRAP marker analyses. Arch Biol Sci. 63(3): 667-679.

Mishra MK, Suresh N, Bhat AM, Surya Prakash N, Satheesh Kumar S, Anil Kumar, Jayarama. 2011b. Genetic molecular analysis of Coffea arabica (Rubiaceae) hybrids using SRAP markers. Rev Biol Tropical. 59(2): 607-617.

Mandal AB, Maiti A, Chowdhury B, Elanchezhian R. 2001. Isoenzyme markers in varietal identification of banana. *In vitro* Cell Dev Biol Plant. 37: 599-604.

Mondal TK, Bhattacharya A, Laxmikumaran M, Ahuja PS. 2004. Recent advances of tea (*Camellia sinensis*) biotechnology. Plant Cell Tiss Org Cult. 76: 195-254.

Moss DW. 1982. Isozymes. Chapman and Hall, London. pp. 204.

Mujib A. 2005. Colchicine induced morphological variants in pineapple. Plant Tissue Cult Biotechnol. 15: 127-133.

Muniswamy B, Kosaraju B, Mishra MK, Yenugula R. 2017. Field performance and genetic fidelity of micropropagated plants of *Coffea canephora* (Pierre ex A. Froehner). Open Life Sci. 12: 1-11.

Muniswamy B, Sreenath HL, Srinivasan CS. 2002. Field performance of tissue cultured plants of *Coffea arabica*. Proceedings of Placrosym XV. pp. 261-265.

Muthiah JVL, Shunmugiah KP, Manikandan R. 2013. Genetic fidelity assessment of encapsulated in vitro tissues of Bacopa monnieri after 6 months of storage by using ISSR and RAPD markers. Turk J Bot. 37: 1008-1017.

Nadha HK, Kumar R, Sharma RK, Anand M, Sood A. 2011. Evaluation of clonal fidelity of *in vitro* raised plants of *Guadua angustifolia* Kunth using DNA-based markers. Journal of Medicinal Plants Research. 5(23): 5636-5641.

Narula A, Kumar S, Srivastava PS. 2007. Genetic fidelity of in vitro regenerants, encapsulation of shoot tips and high diosgenin content in *Dioscorea bulbifera* L., a potential alternative source of diosgenin. Biotechnol Lett. 29: 623-629.

Negi D, Saxena S. 2010. Ascertaining clonal fidelity of tissue culture raised plants of *Bambusa balcooa* Roxb. using inter simple sequence repeat markers. New For. 40(1): 1-8.

Nyende AB, Schittenhelm S, Mix-Wagner G, Greef J. 2003. Production, storability, and regeneration of shoot tips of potato (*Solanum tuberosum* L.) encapsulated in calcium alginate hollow beads. In Vitro Cell Dev Biol Plant. 39: 540-544.

Orbovic V, Calovic M, Viloria Z, Nielsen B, Gmitter Jr FG, Castle WS, Grosser JW. 2008. Analysis of genetic variability in various tissue culture-derived lemon plant populations using RAPD and flow cytometry. Euphytica. 161: 329-335.

Peng J, Kononowiez H, Hodges TK. 1992. Transgenic indica rice plants. Theor Applied Genet. 83: 855-863.

Peng X, Zhang T, Zhang J. 2015. Effect of subculture times on genetic fidelity, endogenous hormone level and pharmaceutical potential of *Tetrastigma hemsleyanum* callus. Plant Cell Tiss Organ Cult. 122: 67-77.

Peyvandi M, Noormohammadi Z, Banihashemi O, Farahani F, Majd A, Hosseini-Mazinani M, Sheidai M. 2009. Molecular analysis of genetic stability in long-term micropropagated shoots of *Olea europaea* L. (cv. Dezful). Asian J Plant Sci. 8:146-152.

Phillips RL, Kaeppler SM, Olhoft P. 1994. Genetic instability of plant tissue cultures: Breakdown of normal controls. Proc Nat Acad Sci USA. 91: 5222-5226.

Phinney B. 1985. Gibberellin A1 dwarfism and shoot elongation inhigher pla nts. Biol Plant. 27: 172-179.

Prasada A, Singh M, Yadav NP, Mathur AK, Mathur A. 2014. Molecular, chemical and biological stability of plants derived from artificial seeds of *Centella asiatica* (L.) Urban- An industrially important medicinal herb. Industrial Crops Products. 60: 205-211.

Rahman MH, Rajora OP. 2001. Microsatellite DNA somaclonal variation in micropropagated trembling aspen (*Populus tremuloides*). Plant Cell Rep. 20: 531-536.

Rani V, Parida A, Raina SN. 1995. Random amplified polymorphic DNA (RAPD) markers for genetic analysis in micropropagated plants of *Populus deltoides* Marsh. Plant Cell Rep. 14: 459-462.

Rani V, Raina SN. 1998. Genetic analysis of enhanced-auxiliary-branching-derived *Eucalyptus tereticornis* Smith and *E. camaldulensis* Dehn Plants. Plant Cell Rep. 17: 236-242.

Rao NK. 2004. Plant genetic resources: advancing conservation and use through biotechnology. Afr J Biotechnol. 3: 136-145.

Ray A, Bhattacharyaa S. 2008. Storage and plant regeneration from encapsulated shoot tips of *Rauvolfia serpentine*- an effective way of conservation and mass propagation. S Afr J Bot. 74: 776-779.

Ray T, Dutta I, Saha P, Das S, Roy SC. 2006. Genetic stability of three economically important micropropagated banana (*Musa* spp.) cultivars of lower Indo-Gangetic plains, as assessed by RAPD and ISSR markers. Plant Cell Tissue Organ Cult. 85: 11-21.

Richardson BJ, Baverstock PR, Adams M. 1986. Allozyme electrophoresis. Academic Press, New York.

Rout GR, Das P, Goel S, Raina SN. 1998. Determination of genetic stability of micropropagated plants of ginger using Random Amplified Polymorphic DNA (RAPD) markers. Bot Bull Acad Sin. 39: 23-27.

Roy B, Mandal AB. 2005. Increased Fe-toxicity tolerance and modulation of isozyme profile in rice. Indian J Biotechnol. 4(1): 65-71

Roy B, Mandal AB. 2008. Development of synthetic seeds involving androgenic embryos and pro-embryos in an elite *indica* rice. Indian J Biotechnol. 7(4): 515-519.

Roy B, Mandal AB. 2011. Profuse microtillering of androgenic plantlets of elite indica rice variety IR 72. *Asian Journal of Biotechnology*, Malaysia. 3(2): 165-176.

Sandoval J, Kerbellec F, Cote F, Doumas P. 1995. Distribution of endogenous gibberellins in dwarf and giant off-types banana (*Musa* AAA cv. 'Grandnain') plants from in vitro propagation. Plant Growth Regul. 17: 219-224.

Scocchi A, Faloci M, Medina R, Olmos S, Mroginski L. 2004. Plantrec overy of cryopreserved apical meristem-tips of *Melia azedarach* L. using encapsulation/ dehydration and assessment of their genetic stability. Euphytica 135: 29-38.

Shah SH, Wainwright SJ, Merret MJ. 2003. Regeneration and somaclonal variation in Medicago sativa and Medicago media. Pak J Biol Sci. 6: 816-820.

Sharma SK, Bryan GJ, WinWeld MO, Millam S. 2007. Stability of potato (Solanum tuberosum L.) plants regenerated via somaticembryos, axillary bud proliferated shoots, microtubers and true potato seeds: a comparative phenotypic, cytogenetic and molecular assessment. Planta. 226: 1449-1458.

Sharma V, Belwal N, Kamal B, Dobriyal AK, Jadon VS. 2016. Assessment of Genetic Fidelity of *in vitro* Raised Plants in *Swertia chirayita* through ISSR, RAPD analysis and Peroxidase Profiling during Organogenesis. Braz Arch Biol Technol. 59: Curitiba 2016 Epub Nov 28.

Shenoy VB, Valil IK. 1995. Biochemical and molecular analysis of plants derived from embryogenic tissue cultures of napier grass (*Pennisetum purpureum* K. Schum). Theor Appl Genet. 83: 947-955.

Singh HP, Ravindranath SD. 1994. Occurrence and distribution of PPO activity in floral organs of some standard and local cultivars of tea. J Sci Food Agri. 64: 117-120.

Singh AK, Sharma M, Varshney R, Agarwal SS, Bansal KC (2006a) Plant regeneration from alginate-encapsulated shoot tips of *Phyllanthus amarus* Schum and Thonn, a medicinally important plant species. In Vitro Cell Dev Biol Plant. 42: 109-113.

Singh AK, Varshney R, Sharma M, Agarwal SS, Bansal KC. 2006b. Regeneration of plants from alginate-encapsulated shoot tips of *Withania somnifera* (L.) Dunal, a medicinally important plant species. J Plant Physiol. 163: 220-223.

Singh R, Srivastava S, Singh S, Sharma M, Mohopatra T, Singh N. 2008. Identification of new microsatellite DNA markers for sugar and related traits in sugarcane. Sugar Tech. 10: 327-333.

Sondahl MR, Baumann TW. 2001. Agronomy II: Developmental and Cell Biology. In: RJ Clarke, OG Vizthum (eds.), Coffee: Recent Development. pp. 202-223.

Sondahl MR, Sondahl CN, Gonclaves W. 1999. Custo comparative de diferentes tecnicas de clonagem. In: III SIBAC Symposium, Londrina, Brazil.

Squirrell J, Hollingsworth PM, Woodhead M, Russell J, Lowe AJ, Gibby M, Powell W. 2003. How much effort is required to isolate nuclear microsatellites from plants? Mol Ecol. 12: 1339-1348.

Srivastava S, Gupta PS. 2008. Inter simple sequence repeat profile as a genetic marker system in sugarcane. Sugar Tech. 10(1): 48-52.

Srivastava V, Khan SA, Banerjee S. 2009. An evaluation of genetic fidelity of encapsulated microshoots of the medicinal plant: *Cineraria maritima* following six months of storage. Plant Cell Tiss Org Cult. 99: 193-198.

Standardi A, Piccioni E. 1998. Recent perspective on synthetic seed technology using nonembryogenic in vitro-derived explants. Int J Plant Sci. 159: 968-978.

Sun D, Li ., Li HB, Ai J, Qin HY, Piao ZY. 2014. Plantlet regeneration through somatic embryogenesis in *Schisandra chinensis* (Turcz) Baill. and analysis of genetic stability of regenerated plants by SRAP markers. Bangladesh J Bot. 44.

Tabassum B, Nasir IA, Farooq AM, Rehman Z, Latif Z, Husnain T. 2010. Viability assessment of *in vitro* produced synthetic seeds of cucumber. African Journal of Biotechnol. 9(42): 7026-7032.

Tautz D, Renz M. 1984. Simple sequences are ubiquitous repetitive components of eukaryotic genomes. Nucleic Acids Res. 12: 4127-4138.

Thomas J, Raj Kumar R, Mandal AKA. 2006. Metabolite profiling and characterization of somaclonal variants in tea (*Camellia* spp.) for identifying productive and quality accession. Phytochemistry. 67: 1136-1142.

Wachira FN, Waugh R, Hackett CA, Powell W. 1995. Detection of genetic diversity in tea (*Camellia sinensis*) using RAPD markers. Genome. 38: 201-210.

Wang Y, Wang F, Zhai H, Liu Q. 2007. Production of a useful mutant by chronic irradiation in sweetpotato. Sci Hortic. 111: 173-78.

Weising K, Nybom H, Wolff K, Kahl G. 2005. DNA fingerprinting in plants: principles, methods, and applications. CRC Press, New York.

Welter L, Gokturk-Baydar N, Akkurt M, Maul E, Eibach R, Topfer R, Zyprian E. 2007. Genetic mapping and localization of quantitative trait loci affecting fungal disease resistance and leaf morphology in grapevine (*Vitis vinifera* L). Mol Breeding. 20: 359-374.

Williams JGK, Kublelik AR, Livak KJ, Rafalski JA, Tingey SV. 1990. DNA polymorphism's amplified by arbitrary primers are useful as genetic markers. Nucleic Acids Res. 18: 6531-6535.

Withers LA. 1983. Germplasm storage in plant biotechnology. In: Mantell SH, Smith H (eds) Plant biotechnology. Cambridge University Press, UK. pp. 187-218.

Xu H, Ton Q, Zhuang W. 1987. Studies on genetic tendency of tea plant hybrid generation using isozyme technique. In: Proc. Int. Tea Quality and Health Symp., Hangzhou, China. pp. 21-25.

Yadav K, Aggarwal A, Singh N. 2013. Evaluation of genetic fidelity among micropropagated plants of *Gloriosa superba* L. using DNA-based markers- a potential medicinal plant. Fitoterapia. 89: 265-270.

Zaefizadeh M, Goliev M. 2009. Diversity and relationships among durum wheat landraces (Subconvars) by SRAP and phenotypic marker polymorphism. Res J Biol Sci. 4: 960- 966.

Zarghami R, Pirseyedi M, Hasrak S, Sardrood BP. 2008. Evaluation of genetic stability in cryopreserved *Solanum tuberosum*. Afr J Biotechnol. 7: 2798-2802.

Zhang M, Wang H, Dong Z, Qi B, Xu K, Liu B. 2010. Tissue cultureinduced variation at simple sequence repeats in sorghum (*Sorghum bicolor* L.) is genotype-dependent and associated with down-regulated expression of a mismatch repair gene, MLH3. Plant Cell Rep. 29: 51-59.

Zietkiewicz E, Rafalski A, Labuda D. 1994. Genome fingerprinting by simple sequence repeat (SSR)-anchored polymerase chain reaction amplification. Genomics. 20: 176-183.

# 10

# Limitations

Results of intensive researches in the field of synthetic seed technology seem promising for propagation of crop plants. However, as described in literature, the major stumbling block in establishing artificial seed production as a viable technology is a lack of understanding of the somatic embryo process and an inability to consistently produce high-quality propagules that can germinate in a soil environment with an acceptably high success rate7. Several aspects of the techniques are still underdeveloped and hinder its commercial application.

## 10.1. LACK OF STANDARD PROTOCOLS

The synthetic seed technology application in a commercial scale is still hampered by the low efficiency of the protocols and logistic behind it. Till today standard protocol for synthetic preparation is available for a limited number of plant species.

## 10.2. HIGH PRODUCTION COST

High hand labour requirements and costly procedures act as barrier for the production of encapsulated propagules at commercial scale and restrict this technology at research level only. Propagation through synthetic seeds needs intensive labour, rooting of regenerated shoots and transplantation, slow and small-scale multiplication. Park *et al.* (1998) developed a cost-effective automated handling system of artificial seed technology. They developed this technology to produce high value clones in clonal forestry and it can be achieved cheaply by mass serial rooting of cuttings from juvenile donor plants produced from cryopreserved embryogenic cultures. Sicurani *et al.* (2001) also produced synthetic seeds from machine-ground explants. Harrell *et al.* (1992) developed a fluidic, ***in vitro*** embryo monitor and harvester for fluid based, bioreactor culture systems. A machine vision system had been used to classify embryos as they were pumped through an in-line imaging cell. Embryo morphologies were distinguished using a battery of geometric features extracted by the vision system and a statistical based decision rule or optionally with a neural network. The position and velocity of embryos existing the imaging cell were tracked with an array of photodiodes, the time required to execute image-processing algorithms. Embryos, which confirmed to a targeted morphology, were routed into a harvest port with a precisely timed injection of culture medium. Objects which were not harvested were recycled back into the bioreactor. The system was capable of a maximum harvest rate of one particle per second. Hroslef-Eide *et al.* (2003) also suggested the mechanical approach for somatic embryo production to decrease the labour cost.

## 10.3. STRAIN IN RECURRENT PRODUCTION OF SOMATIC EMBRYOS

Although large quantities of somatic embryos can be rapidly produced in many plant species, continuous supply is difficult as the totipotency decrease with age of culture. Parrot and Bailey (1993) standardized protocol for recurrent supply of somatic embryos of alfalfa. They reported that new somatic embryos form repeatedly directly on older somatic embryos when cultured on growth regulator free MS (Murashige and Skoog, 1962) medium supplemented with B5 (Gamborg, 1972) vitamins. Those cultures were maintained for 2 years, during which time their embryogenic capacity remained stable. Roy and Mandal (2006) also developed tissue cultural technique to produce rapid and recurrent androgenic microtiller of rice.

## 10.4. ASYNCHRONOUS DEVELOPMENTOF SOMATIC EMBRYOS

However, because of certain inherent problems, the rate of production of uniform and high quality embryos is much lower as a result of which the preparation of efficient and quality seeds has been successful in only a few crop plants (Datta and Potrykus, 1989).Improper maturation of the somatic embryos and asynchronous development is a problem in artificial seed preparation. Development of artificial seeds requires sufficient control of somatic embryogeny from the explants to embryo production, embryos development and their maturation. Hence, mature somatic embryos must be capable of germinating out of the capsule or coating to form vigorous normal plants. In some cases, somatic embryos often develop extra cotyledons or poorly developed apical meristems (Ammirato, 1987). This asynchronous embryo development makes harvest difficult. Uniformly mature somatic embryo development have included physical separation of proembryonal cultures to assure uniform callus size and physiological synchronization by adding abscisic acid appears to cause cell water (turgor) content to decrease, thereby slowing embryo growth which inhibit germination of embryos that would tend to germinate precociously (Gray, 1989). A number of researchers have tried to improve the quality and quantity of somatic embryos in modification of culture conditions (Kantharajah and Golegaonkar, 2004), such as, medium composition (Ipekci and Gozukirmizi, 2003; Cheeand Canteliffe, 1992), growth regulators (Ipekci and Gozukirmizi, 2003; Lata *et al.,* 2004), physical state of the medium (Senaratna *et al.,* 1995). Asynchronous maturation is also a strain in synthetic seed production. Attree and Fowke (1993) have described that inclusion of high levels of sucrose in the standard medium containing ABA, prevents maturation, while inclusion of PEG with ABA dramatically improves the frequency and synchrony of the somatic embryo maturation.

Madakadze and Senaratna (2000) investigated the interactions of growth regulators and polyethylene glycol (PEG) on the maturation of *Pelargonium hortorum***.** They found that BAP and NAA did not enhance ABA effects on maturation frequency but only improved frequency in the presence of polyethylene glycol. ABA significantly improved protein content in the presence of PEG. BA and NAA in combination with ABA and jasmonic acid improved protein types in somatic embryos only in the absence of PEG. Osmoticum caused by polyethylene glycol was the main component required for protein synthesis.

## 10.5. MULTIPLE SOMATIC EMBRYOS DEVELOPMENT

Multiple somatic embryos are often found on a single callus, in which multiple stages of embryo development are observed. This causes the non-uniform embryos to be subjected to change nutrient conditions since the nutrients are depleted by the developing tissues and the replenished. Consequently many somatic embryos have organs developing at different rates, which contribute to asynchronous embryo development (Conger *et al.*, 1989). In some cases, this leads to precocious germination, while in others the prevailing nutrient environment may be conducive to shoot or root development but not both.

## 10.6. POOR CONVERSION OF ARTIFICIAL SEED INTO PLANTLETS

For commercial applications, somatic embryos must germinate rapidly and should be able to develop into plants at least at a rate and frequencies more or less similar, if not superior to true seeds. The germination and plantlet conversion ***in vitro*** is considerably high enough, but which fail to provide as many plantlets in soils. The low germination and conversion capacity of synthetic seeds may be due to the absence of a nutritive tissue like the endosperm of the natural seed (Sanatha *et al.*, 1993) and due to this hindrance of the gel capsule for the emergence of the root and shoot. Providing a artificial endosperm and adopting the self-breaking alginate gel beads technology could overcome these shortcomings (Onishi *et al.*, 1994). Restricted oxygen diffusion, increased ethylene production and hindrance of coating material for shoot and radical emergence have been reported to have drastic effects on germination of synthetic seed (Kursulee *et al.*, 1992). Adoption of self-breaking synthetic seed technology significantly enhanced germination by 50% and conversion by 40% in comparison to normal synthetic seeds (Arum Kumar *et al.*, 2005).

To achieve conversion of somatic embryos into plantlets and to overcome deleterious effects of recurrent somatic embryogenesis as well as abnormal development of somatic embryos on their conversion, it is necessary to provide optimum nutritive and environmental conditions (Capuano and Debergh, 1997). Maltose has been found valuable for improving alfalfa somatic embryo conversion (Redenbaug and Walker, 1990). From a synthetic seed perspective, addition of sucrose in the medium is necessary for viability of somatic embryos, their subsequent development, maturation and germination in many plant species (Jain *et al.,* 1995).

## 10.7. LACK OF DORMANCY

Somatic embryos do not typically exhibit a quite resting and dormancy (quiescence) phase characteristic of *orthodox* seeds (Gray, 1987a, b). Usually somatic embryos continues to grow into seedling or they revert back into disorganized callus tissue. This inability to produce a resting phase where all embryos are at the same arrested physiological and morphological state also is a challenge to synthetic seed development. Without this arrested growth stage, synthetic seed cannot be successfully stored or treated using traditional seed technology practices. Attempts to introduce dormancy into somatic embryos are being explored using orthodox seeds as a developmental model. Since orthodox seed expressed dormancy after being dehydrated on the parent plant, studies have been initiated to determine whether they dry down or somatic embryos might induce this important trait and there is evedience of success (McKersie *et al.*, 1988; Carman, 1988; Senaratna *et al.*, 1989).

## 10.8. LACK OF STRESS TOLERANCE

In many plant species, the somatic embryos have been found to be sensitive to desiccation. Desiccation damages the somatic embryos and inhibits their germination and conversion into plants in desiccation sensitive species (Redenbaugh, 1993). Gradual drying of alfalfa somatic embryos with progressive and linear loss of water gave better response and improved the quality of embryos in comparison to uncontrol drying (Redenbaugh and Walker, 1990). Senaratna *et al.* (1995) have reported that desiccation tolerance can be induced in somatic embryos of alfalfa by external stimuli such as ABA, exposure to cold, heat and osmatic stress at sub-lethal levels or increasing the sucrose content in the medium. PEG and ABA were very tolerant to low moisture levels. Such somatic embryos had less than 50% moisture content, which was further reduced to less than 10% following desiccation. These embryos were stored at –20 °C for a year and thereafter successfully germinated following inhibition with no loss in viability.

## 10.9. LACK OF GENETIC FIDELITY

Genetic variability for regeneration via somatic embryogenesis has been documented for a wide variety of species (Parrot *et al.*, 1989), maize (Hodges *et al.*, 1986), rice (Abe and Futsuhara, 1986), barley (Ohkoshi *et al.*, 1987), wheat (He *et al.*, 1988). Genetic control of regeneration capacity is largely additive and highly heritable.

## 10.10. PROBLEMS ASSOCIATED WITH GELLING AGENTS

The coating materials may also limit success of synthetic seed technology. The hydrated capsules are more difficult to store because of the requirement of embryo respiration. A second problem is that capsules dry out quickly unless kept in a humid environment or coated with a hydrophobic membrane. Calcium alginate capsules are also difficult to handle, because they are very wet and tend to stick together slightly. In addition, calcium alginate capsules lose water rapidly and dry down to a hard pellet within a few hours when exposed to the ambient atmosphere. Redenbaugh *et al.* (1993) have reported that the limitations caused by coating materials can be overcome by selecting appropriate coating material for encapsulation. The coating material should be non-damaging to the embryo, mild enough to protocol, the embryo and allow germination and should be sufficiently durable for rough handling during preparation, storage, transportation and planting. The coat must contain nutrients, growth regulator(s) and other components necessary for germination and conversion.

## 10.11. PROBLEMS ASSOCIATED WITH TREE SPECIES

Tree species may exhibit a unique set of tissue culture-dependant varieties. Genetic variation may be higher for tree species than cultivated herbaceous species. The size of the genome is quite large for many trees, which is particularly important for genetic engineering. A further limitation for tree improvement using genetic engineering is that mature material must be collected from the field rather than controlled environment. In addition, the problem of ecotype or physiological variation from individual to individual is much greater due to the overall heterozygosity of most tree species. This additional variation further increases the potential difficulty of developing repeatable somatic embryogenesis in tree species.

## References


Abe T, Futsuhara Y. 1986. Genotype variability for callus formation and regeneration in rice (*Oryza sativa* L.). Theor Appl Genet. 72: 3-10.

Ammirato PV. 1987. Organizational events during somatic embryogenesis. In: Plant Tissue and Cell Culture, New York, Alan Liss.

Arun Kumar MB, Vakeswaran V, Krishnasamy V. 2005. Enhancement of synthetic seed conversion to seedlings in hybrid rice. Plant Cell Tiss Org. 81: 97-100.

Attree SM, Fowke LC. 1993. Embryogeny of gymnosperms: Advances in synthetic seed technology of conifers. Plant Cell Tiss. Org. Cult. 35 (1): 1-35.

Capuano M, Debergh PC. 1997. Improvement of maturation and germination of horse chestnut somatic EMBRYOS. Plant Cell Tiss Org Cult. 48(1): 23-29.

Carman JG. 1988. Improve somatic embryogenesis in wheat by partial simulation of the *inovulo* oxygen growth regulator and desiccation environment. Planta. 175: 417-422.

Chee RP, Cantliffe DJ. 1992. Improved production procedures for somatic embryos of sweet potato for a synthetic seed system. Hort Science. 27(12): 1314-1316.

Conger BV, Hovanesian JC, Trigiano RN, Gray DJ. 1989. Somatic embryo ontogeny in suspension culture of orchard grass. Crop Sci. 29: 448-452.

Datta SK, Potrykus I. 1989. Artificial Seeds in Barley: Encapsulation of Microspore-Derived Embryos. TheorAppl Genet. 77: 820-824.

Gray DJ. 1987a. Effects on hydration and other environmental factors on dormancy in grape somatic embryos. Hortl Sci. 23: 786-791.

Gray DJ. 1987b. Quiescence in monocotyledonous and dicotyledonous somatic embryos induced by dehydration.Hort Sci. 22: 810-814.

Gray DJ. 1989. Synthetic seeds for clonal production of crop plants. In: Recent Advances in the Development and Germination of Seeds, Taylorson RB (ed.), New York, Plenum Press. pp. 29-45.

Harrell RC, Hood CF, Manilla RD, Molto EC, Beienik M, Cantliffe DJ. 1992. Automatic identification and separation of somatic embryos ***in vitro***. International Symposium Transplant Production Systems. Proceedings of a conference held in Yokohama, Japan, 21-26 July, 1992, p. 90.

He DG, Yang YM, Scott KJ. 1988. A comparison of scutellum callus, embryo age and medium. Plant Sci. 57: 225-233.

Hodges TK, Kamo KK, Imbrie CW, Becwar MR. 1986. Genotype specificity of somatic embryogenesis and plantlet regeneration of maize.Biotechnol. 4: 218-223.

Hroslef-Eide AK, Munster C, Heyerdahl PH, Lyngved R, Olsen OAS, Saxena P. 2003.Liquid culture systems for plant propagation.ActaHorticulturae. 625: 173-183.

Ipekci Z, Gozukirmizi N. 2003. Direct somatic embryogenesis and synthetic seed production from *Paulownia elongate*. Plant Cell Rep. 22(1): 16-24.

Jain SM, Gupta PK, Netwon RJ. 1995. Somatic embryogenesis in woody plants, Kluwer Academic Publishers, Dordrecht.

Kantharajah AS, Golegaonkar PG. 2004. Somatic embryogenesis in egg plant. ScientiaHorticulturae. 99(2): 107-117.

Kursulee A, Corbinau F, Hervagault JF. 1992. Limitation of oxygen diffusion to carrot somatic embryos by encapsultion. In: Proccedings of Fourth Workshop on Seed, Corbineau CD (ed.). Mac Millan Press, New York. pp. 193-199.

Lata H, Bedir E, Moraes RM, Andrade Z, Lewinsohn E. 2004. Mass propagation of *Echinoceaeangustifolia* : a protocol refinement using shoot encapsulation and temporary immersion liquid system. ActaHorticulturae. 629: 409-414.

Madakadze RM, Senaratna T. 2000. Effect of growth regulators on maturation of gerenium (*Pelargonium hortorum*) somatic embryos. Plant Growth Regulation. 30(1): 55-60.

McKersie BD, Bowley SR, Senaratna T, Brown DCW, Bewley JD. 1988. Application of artificial seed technology in the production of alfalfa (*Medicago sativa* L.). *In Vitro* Cell Dev Biol. 24: 2171-2178.

Murashige T and Skoog F. 1962. A Revised medium for rapid growth with tobacco tissue culture. Physical Plant. 15: 473-497.

Ohkoshi S, Komatsuda T, Enomoto S, Taniguchi M, Ohyama K. 1987. Use of tissue culture for barley improvement: I. Differences in callus formation and plantlet regeneration from immature embryos of 179 varieties. Japan J Breed. 37: (Suppl. 1): 42-43.

Onishi N, Sakamoto Y, Hirosawa T. 1994. Synthetic seeds as an application of mass production of somatic embryos. Plant Cell Tissue Org Cult. 39(2): 137-145.

Park YS, Barrett JD, Bonga JM. 1998. Application of somatic embryogenesis in high-value clonal forestry: development, genetic control, and stability of cryopreserved clones. *In Vitro* Cellular DevlBiol Plant. 34(3): 231-239.

Parrot WA, Bailey MA. 1993. Characterization of recurrent somatic embryogenesis of alfalfa on auxin-free medium. Plant Cell Tiss Org Cul. 32(1): 69-76.

Parrot WA, Williams EG, Hildebrand DF, Collins GB. 1989. Effects of genotype on somatic embryogenesis from immature cotyledons of soybean. Plant Cell Tiss Org Cult. 16: 15-21.

Pond S, Cameron S. 2003. Artificial Seeds, Tissue Culture, Elsevier Ltd. pp. 1379-1388.

Redenbaugh K, 1993. Synseeds: Application of synthetic seeds to crop improvement. CRC Press, Boca Raton.

Redenbaugh K, Fujii JA, Slade D. 1993. Hydrated coatings for synthetic seeds. In: Synseeds: application of synthetic seeds to crop improvement, Redenbaugh K (ed.). CRC, Boca Raton. pp. 35-46.

Redenbaugh K, Walker K. 1990. Role of artificial seeds in alfalfa breeding. In: Plant tissue culture: application and limitations. Development in crop science, Bhojwani SS (ed.). 19-Elsevier, Amsterdam, pp. 102-135.

Roy B, Mandal AB. 2006. Rapid and recurrent mass-multiplication of androgenic embryos in *indica* rice. Indian J Biotechnol. 5(2): 239-242.

Sanatha M, Sakamoto Y, Hayashi M, Mashiko T, Okamoto A, Onishi N. 1993. Clerey and lettuce. In: Synseeds, Redenbaugh K (ed.). CRC Press, Boca Raton. pp. 305-327.

Senaratna T, Saxena PK, Rao MV, John Afele. 1995. Significance of *Medicago sativa* L. somatic embryos. Plant Cell Rep. 14: 375-379.

Sicurani M, Piccioni E, Standardi A. 2001. Micro-propagation preparation of synthetic seed in M.26 apple root stock I: attempts towards saving labour in the production of adventitious shoot tips suitable for encapsulation. Plant Cell Tiss Org Cult. 66(3): 207-216.

# 11

# Future Scenarios

Micropropagation through artificial seeds may be commercially exploited on a large scale, generating millions of plantlets within a few days, and this may become a profitable multibillion rupees industry in near future. This technology would be feasible and even competitive economically with the other seed production methods.

In the near future, the synthetic seed technology could be the key to overcome these problems, e.g., through automated procedures of artificial seed production (Aitken-Christie *et al.* 1995) followed by autotrophic micropropagation (Jeong *et al.* 1995). The synthetic seed systems coupled with artificial intelligence and microcomputer systems like the most advanced robots, which can mimic the motions, and functions of living beings (i.e. automated encapsulations). It would tremendously increase the efficiency of encapsulation and production of artificial seeds, and revolutionized the plant propagation method in the years ahead. This technology is gradually moving towards the commercial propagation of high value crops. However, there is a great need for refinement of this technology by tackling of certain technical problems such as the need to produce high quality and high fidelity somatic embryos and to avoid the genetic instability and variability of tissue culture derived plants. The somatic variation can be overcome if the process and regulation of somatic embryogenesis and origin of somaclonal variation are well understood. The understanding about the storage, transport, handling growth habit and harvest index of artificial seeds is essential. The efforts are also needed to increase the output of embryos or plants per gram of callus tissue and conversion frequency of artificial seeds are also needed.

Somatic embryogenesis has been achieved for many species; however, for many of the important high value genotypes, somatic embryogenesis has not been reported. Therefore, it needs to shift the focus on somatic embryogenesis in valuable crop plants. Alternative coating materials for encapsulating embryo and other vegetative propagules are in urgent need to increase conversion rate.

Although a large number of reports are available on successful production and subsequently plantlet conversion of synthetic seeds (Table 7.1), practical implication of the technology is constrainted due to several major problems, which hinder its commercialization. The first requirement for the practical application of synthetic seed technology is large-scale production of high quality micropropagules, which is at present a limiting factor. The second limiting factor is the survival and germination of the encapsulated somatic embryos is low. Poor germination of synthetic seeds may be due to lack of nutrient and/or oxygen supply, microbial infection and mechanical damage of the embryo. The conversion of desiccated somatic embryos to plantlets is low. The desiccation process, which damages the embryos, and other problems associated with desiccated synthetic seeds need regulation. Third major obstacle in synthetic seed technology is reduction in viability on storage. Most of the research efforts reflect on short-term storage only. Further research is needed for prolong maintenance viability of somatic embryos. Either hydrated calcium alginate based or desiccated polyoxyethylene glycol based synthetic seeds might be used, but it is likely that some degree of drying before cryopresevation would be beneficial. Occurrence of somaclonal variation in tissue culture is another aspect to be considered which recommends the use of synthetic seeds for clonal propagation. In nutshell, somatic seed remains a poor analog of natural seed in terms of viability, handling and storage. More basic comparative biochemical and physiological research is needed to understand the differences in response between zygotic and somatic seed, so as to determine whether or not somatic embryo is behaving like recalcitrant seed and/or like an isolated zygotic embryo *in vitro*.

Synthetic seed technology provides a rich analog of natural seed in terms of clonal propagation, handling in national and international exchange of seeds for genetic improvement and germplasm conservation.

## References

Aitken-Christie J, Kozai T, Smith MAL. 1995. Glossary. In: Aitken-Christie J, Kozai T, Smith MAL (eds) *Automation and Environmental Control in Plant Tissue Culture*, Kluwer, Dordrecht, pp ix-xii.

Jeong RB, Fujiwara K, Kozai T. 1995. Environmental control and photoautotropic micro-propagation. In: Janick J (ed) *Horticultural Reviews* (vol 17), Wiley, NY. pp. 125-172.

# Subject Index

## D



## E

## T



## U



## V

## W



## H



## Z